Open Seawater

Public, Professionals and Preservation:
In Situ Conservation of Cultural Heritage

Edited by:
Vicki Richards
Jennifer McKinnon

Seabed

ALL PAPERS IN THIS PUBLICATION HAVE UNDERGONE A PEER-REVIEW PROCESS.
FRONT COVER IMAGES COURTESY OF DAVID GREGORY.

A PAST FOUNDATION PUBLICATION
WWW.PASTFOUNDATION.ORG

Foreword

In September 2008, a combined annual conference for the Australasian Institute for Maritime Archaeology, the Australasian Society for Historical Archaeology and the Australian Association for Maritime History was held at the magnificent Institute Building of the State Library of South Australia in Adelaide. The conference was entitled "Archaeology from Below – Engaging the Public" and hoped to address the relationship between archaeology and the public. The conference attracted a significant number of professionals, avocationals and interested members of the public with attendance reaching just over 150 individuals. Participants came from a diverse range of backgrounds but the one common denominator was their passion and interest in our cultural heritage.

Over the three days of the conference, there were six plenary lectures presented by six international guest speakers and nine conference sessions featuring fifty presentations in total. One of the largest sessions, with a total of eleven papers, was "Public, Professionals and Preservation: Conservation of Cultural Heritage." The aim of this session was to broaden discussions on the interactions of professionals and the public with respect to *in situ* stabilisation, preservation and management of terrestrial and underwater cultural heritage sites and their associated archaeological finds.

Papers presented in this session discussed this basic theme, including the ideology of on-site preservation, legislative requirements, present directions of in-situ preservation, assessment of site and artefact deterioration, principles for the development and implementation of on-site mitigation strategies, long-term effects of past stabilisation techniques and *in situ* monitoring of sites to determine the effectiveness of management strategies. These papers were delivered by a range of professionals including conservation scientists, students, consultants, maritime archaeologists and government practitioners. Interest in *in situ* preservation has been increasing exponentially since 2000 but for Australia this is the first time that such a large collection of papers on the topic of *in situ* stabilisation, preservation and management by a wide range of practitioners has been presented at an archaeology conference.

That this is such a rare occurrence in Australia and around the globe demonstrates the divide between those conducting research into *in situ* stabilisation and preservation (conservation scientists) and those attempting to apply *in situ* management techniques (archaeologists). This divide has proven to be detrimental to the progress of *in situ* management and likely originates from our lack of communication. A quick review of *in situ* stabilisation and preservation literature sends you to journals such as *International Biodeterioration and Biodegradation* and *Trends in Corrosion Research* or to conference proceedings including *Proceedings of the ICOM Group on Wet Organic Archaeological Materials Conference* or *Proceedings of the International Conference on Metals Conservation*. On occasion we may even get lucky and connect in the *Journal of Archaeological Science*, but those occasions are uncommon. This is not to put the blame on conservation scientists for publishing in their own journals and attending their own conferences, nor is it to accuse archaeologists of not reading widely enough or finding comfort in attending more papers on shipwreck finds. Rather it is a call for both to interact with regularity and conviction.

One project, which has included both archaeologists and conservation scientists working together from the planning process is outlined in Carpenter, MacLeod and Richards' paper, *Conserving the WWII Wrecks of Truk Lagoon*. Carpenter et al. outline a project directed by Dr Bill Jeffery, which considers conservation a key element to the study of WWII sites. By outlining a conservation strategy early, conservation scientists like Carpenter, MacLeod and Richards can provide critical data for the management of such sites.

One of the biggest critiques from our antagonists, those who profit from submerged heritage (i.e. treasure hunters), is that we practitioners use *in situ* as a means to "do nothing." Perhaps this misstatement stems from an ignorance on both archaeologists and profiteers about the definition of *in situ* preservation and management. In *Developing Methodology for Understanding In Situ Preservation and Storage from a Practitioner Perspective*, Ortmann attempts to address this issue with a definition of *in situ* preservation: "Any steps taken on or intervention with a site in order to extend its longevity while maintaining original context and spatial position; while artefacts and features may have been excavated and/or removed, the site itself remains in place and retains all or a majority of its original context." By her very definition Ortmann suggests that "any steps," be it site monitoring or full-scale reburial, constitutes *in situ* preservation.

Gregory in his paper *In Situ Preservation of Marine Archaeological Sites: Out of Site but Not Out of Mind*, further defines the method of *in situ* preservation and management by outlining "five fundamental steps to ensure the successful and responsible *in situ* preservation of underwater archaeological sites." His article is a must-read for practitioners as it underscores the fact that *in situ* preservation "is not a new idea and has been practiced over the past 20-30 years."

Cegrass™, Sand and Marine Habitats: A Sustainable Future for the William Salthouse *Wreck* by Steyne is a paper on just one of the many successful programs of *in situ* management in Australia. In the paper, Steyne outlines the steady progress, and sometime regress, of the efforts to preserve *in situ* one of Victoria's oldest shipwrecks. The process was one of successes and failures and is still ongoing as a result of recent channel deepening efforts. This example demonstrates that archaeologists and managers are not just yelling *in situ* from their soapbox but are actively "doing."

Muir and Loo in *Reaching Out to the Community: Bringing Leslie and Ross Back Home to Harcourt* discuss a "dry" example of *in situ* exhibition and interpretation. This project is a fantastic example of how government, consultancy and community can work together to keep heritage local. By returning collections to their original communities people can reconnect with their archaeological past thus reducing the need for storage space in repositories.

In times of decreasing state and federal funding, *in situ* preservation is becoming a major topic in archaeological heritage management discussions. In addition, due to the increasing costs of excavation, long-term storage, specialised conservation treatments and post conservation display and storage of recovered cultural material, it is not unexpected that the archaeological community and heritage managers are moving away from the more traditional excavation methods and further towards *in situ* management of our cultural heritage. Raupp, Coroneos and McKinnon's article, *Excavation and Relocation of the Former Hovell Pile Light*, is perhaps an expression of this prudence during a recent mitigation process. The complete excavation, relocation and reburial of the former Hovell Pile Light site is rare in Australia and signals a significant shift in the mind set of thinking *in situ* is the "only" alternative for the preservation of a site.

Along the lines of thrift and ingenuity we have a number of countries including Australia disposing decommissioned warships in the name of artificial reefs for dive tourism along our coastlines. Richards, MacLeod and Morrison's article, *Corrosion Monitoring and the Environmental Impact of Decommissioned Naval Vessels as Artificial Reefs* expresses the importance of monitoring these sites for important information on the synergistic interactions between modern shipwreck materials and the marine environment. Imagine what we may learn from a long-term project such as this about metal ship deterioration and their effects on the marine environment. We may not live to see the day, but our successors will.

In situ preservation has recently been politically galvanised in the ICOMOS Charter for the Protection and Management of the Archaeological Heritage 1990 and the recently ratified UNESCO Convention on the Protection of the Underwater Cultural Heritage 2001. Hence, it is not surprising that this particular session attracted a considerable number of relevant and interesting papers.

The conference session led to some interesting discussions about how we as conservation scientists, students, consultants, maritime archaeologists and government practitioners should collaborate more often. Eventually these discussions led to the notion of publishing these particular papers in a proceedings publication, by which we could showcase the increasing body of knowledge of *in situ* preservation strategies and the growing interest in the *in situ* management of archaeological sites.

We hope that you enjoy reading this publication and that it stimulates you as much as the conference session inspired us to think to think about the concepts of *in situ* stabilisation, preservation and management.

Finally, we would like to offer our thanks to those who contributed to this publication.

Vicki Richards
Department of Materials Conservation,
Western Australian Museum,
Shipwreck Galleries,
45-47 Cliff Street,
Fremantle, Western Australia, 6160

Jennifer McKinnon
Department of Archaeology,
Flinders University,
GPO Box 2100,
Adelaide, South Australia, 5001

Table of Contents

List of Figures

Figures Continued

List of Tables

1 *In situ* Preservation of Marine Archaeological Sites: Out of Sight but Not Out of Mind

David Gregory
The National Museum of Denmark, Conservation Department, I.C. Modewegsvej, Brede, DK-2800, Kongens Lyngby, Denmark

In situ preservation is increasingly being seen as a means to manage marine archaeological sites which, for economic reasons and current international trends favouring *in situ* preservation, are not excavated, raised and conserved. However, *in situ* preservation should not be a case of leaving a site where it is – out of sight, out of mind - and hoping that it will be there when archaeologists and conservators have the capacity, research questions and desire to investigate these finds, in the future. Five fundamental steps to ensure the successful and responsible *in situ* preservation of underwater archaeological sites are discussed:
1. The extent of the site to be preserved
2. The most significant physical, chemical and biological threats to the site
3. The types of materials present on the site and their state of preservation
4. Strategies to mitigate deterioration and stabilise the site from natural impacts
5. Subsequent monitoring of a site and implemented mitigation strategies
An overview of research concerning wooden wrecks addressing these five points will briefly be presented in the paper.

Introduction

On the 2nd of January 2009 the UNESCO Convention for the Protection of the Underwater Cultural Heritage (http://unesdoc.unesco.org/images/0012/001 260/126065e.pdf) came into force. The need for such a convention arose, " *first and foremost...to gain acceptance of the idea that the underwater cultural heritage is part of the universal heritage of humanity, just as significant and deserving the same protection as the cultural heritage found on dry land...*" Furthermore, "*the Convention and its Annex are based above all upon the elimination of the law of salvage and preventing the commercial exploitation of the underwater cultural heritage*" (Grenier 2006). In terms of affording protection to the underwater resource, and of particular importance to the current paper, are Paragraphs 5 and 10 of Article 2 - Objectives and general principles. These state that as a first option the underwater cultural heritage should be protected *in situ* and where possible advocate the use of non-intrusive methods to document and study these sites *in situ*. Thus the convention sets a political framework for the *in situ* protection of the submerged cultural heritage

Nonetheless, concerns have been expressed toward this policy; "*preservation in-situ has been likened to watching the archaeology rot with an expensive programme of monitoring. It is claimed that the ability to preserve sites in-situ has not currently be proven yet and that more scientific research on the practicalities of such a strategy still need further investigation.*" (Bulletin of The Maritime Affairs Group, Institute of Field Archaeologists, Winter 2008).

In situ preservation of underwater cultural heritage, or the re-deposition (re-burial) of excavated material into benign environments conducive to its long term preservation, is not a new idea and has been practiced over the past 20-30 years. With a few notable exceptions, it has primarily been a pragmatic solution for the immediate protection of a site following its exposure due to natural causes, to stabilise a site after its partial archaeological excavation or for the long term storage of finds when resources are not available for conservation and curation.

In situ preservation is just one of the tools available to archaeologists, conservators and cultural resource managers when faced with the discovery of a new site or managing existing sites. In this paper a process based approach to understanding both the site environment and the processes of deterioration and *in situ* preservation of wood are considered, although the same fundamental principles can be applied to metal artefacts and wrecks (see Richards et al. this volume). In this approach the following five points are discussed:

1. The extent of the site to be preserved
2. The most significant physical, chemical and biological threats to the site
3. The types of materials present on the site and their state of preservation
4. Strategies to mitigate deterioration and stabilise the site from natural impacts
5. Subsequent monitoring of a site and implemented mitigation strategies

1. Extent of the site to be preserved

The romantic view of a shipwreck is typically an almost intact ship sitting on the sea bed with its sails still attached to the rigging. In reality, what constitutes a shipwreck site are those parts and materials which have survived the wrecking process and, following a period of deterioration/stabilisation, reached an equilibrium with its new environment. However, reaching this equilibrium may, depending on the prevailing environment, mean that the ship and its associated contents do not remain conveniently contained within the original vessel. Furthermore, the equilibrium established is not steady state; it is dynamic (Quinn 2006; Ward et al. 1999) and as a result the status

quo can break down leading to periods of renewed and continued deterioration.

Archaeological investigations understandably focus on examining the structure of the remaining hull and associated finds. However, in terms of managing a site *in situ* is it enough to focus on these immediate finds? Does other archaeological material from the wrecking process or deterioration of the shipwreck remain exposed, or buried, away from the hull itself?

Underwater visibility and the potential scale of a site can often make it difficult for a diving archaeologist to rapidly get a complete "overview" of a site and it can take many dives to completely come to grips with the overall extent of a site and the potential changing nature of a site due to ongoing formation processes. This is even more apparent when it comes to those parts of the assemblage which lie buried within sediments. However, with marine geophysics being increasingly applied on marine archaeological sites, it is now possible to more rapidly assess the spatial distribution of material both on and within the seabed. Notable examples are the work of Quinn et al. (1997, 1998) on the wrecks of *Mary Rose* and HMS *Invincible*. In the case of *Mary Rose*, the area where the hull was found was surveyed using a sub bottom profiler (Chirp) in part to investigate the site for remaining wreck structure. The resulting chirp profiles of the excavated site showed structures which were previously unknown on the site and were interpreted as in-filled scour features associated with the sinking and subsequent degradation of *Mary Rose* in a dynamic tidal environment.

Subsequent archaeological investigation of these in-filled scour features yielded parts of the bow structure of the ship. The reason for this was that as the ship lay on the seabed and was exposed to tidal currents, scour occurred. Quinn (2006) discusses the role of scour in shipwreck site formation processes. Summarily, scour occurs at the seafloor when sediment is eroded from an area in response to force by waves and currents and is often initiated by the introduction of an object (e.g. shipwreck) to the seafloor. As a result scour hollows or pits can be formed and in the right conditions, as with the *Mary Rose*, can serve as a sink for archaeological material removed from the ship during its deterioration. A further use of marine geophysics to assess the extent of a wreck site is the case of HMS *Invincible*. Chirp sub bottom profiler and side scan sonar data gathered over the site showed that the amount of buried archaeology far exceeded what was actually visible on the seabed (Quinn et al. 1998:137).

Both these examples serve to show how a site, defined by what is visible at a given point in time, may actually only constitute a small part of the archaeology present. The application of marine geophysics can quantify the extent of what is above and below the seabed. Furthermore, analysis of seabed formations (e.g. ripple marks, sand waves) obtained through side scan (Anthony and Leth 2002) or multi beam echo sounder (Quinn 2006) can also give an indication of the sedimentary processes ongoing on a site and provide information on the possibility of future sediment erosion or accretion over a site. These factors are equally important when designing a management strategy to protect sites from future deterioration.

2. The most significant physical, chemical and biological threats to the site

Having identified the extent of a site it is necessary to understand the nature of the materials present on the site in terms of their state of preservation and what factors can lead to their further deterioration. Figure 1-1 shows an idealised view of a wooden shipwreck as it may appear after the

wrecking process. Effectively the wreck and its component parts will be exposed to two very differing environments – the open sea water and the sediments of the sea bed.

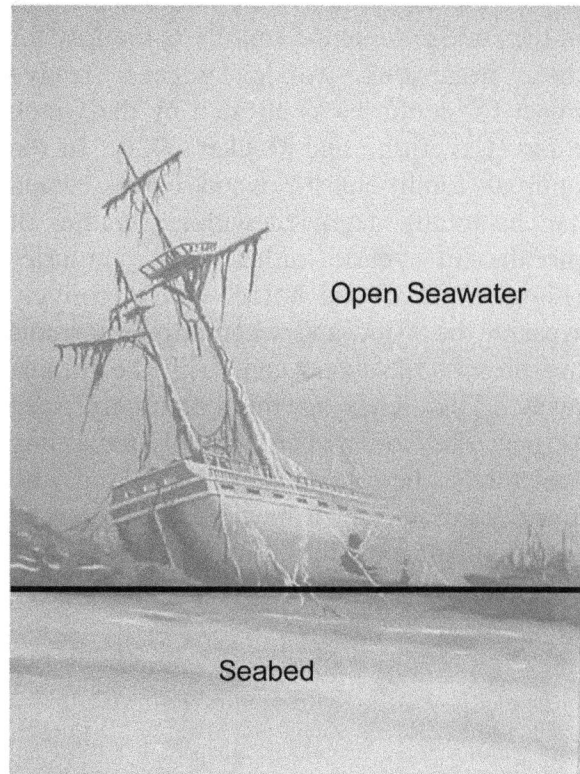

Figure 1-1 An idealised shipwreck on the seabed showing exposed and buried areas (Author).

In the open seawater, physical processes, namely scour and biological processes, are the major causes of deterioration of wooden and organic materials. Quinn (2006) has summarised the role of scour in the formation of a wreck site and, when in conjunction with wood boring organisms, it can lead to the relatively rapid deterioration of those upper parts of a wreck which are not covered by sediment during the initial wrecking phase. Shipworm (wood boring molluscs typically members of the family Teredinidae) or Gribble (wood boring crustaceans typically of the families Isopoda or Amphipoda) require dissolved oxygen from the surrounding seawater in order to respire and in this respect this is the "limiting factor" for their presence on wreck

3

sites. However, salinity and temperature are also key factors which affect their survival, rate of reproduction and rate of activity. (Turner and Johnson 1971; Becker 1971). Where these parameters are not met, such as in the Baltic where the salinity is too low for these organisms, wooden wrecks survive intact for centuries as attested by the wreck *Vasa* (Cederlund and Hocker 2006). In the optimal conditions for wood borers, wood can be totally degraded within a matter of months or years, rather than centuries (Figure 1-2). Wood borers will effectively weaken the wood and when strong currents are present the wood can easily be broken away. This leads to the commonly seen "kipper skeleton"; where the only remaining structure is the bottom runs of planking with some framing, which have been covered by sediment since the initial wrecking.

Figure 1-2 A modern sample of pine wood placed on a wreck in Denmark showing complete deterioration by shipworm (*Teredo spp.*) after just six months (Author).

However, should the wreck come to be covered by sediment, as a result of sediment accretion rather than erosion, or through the process of liquefaction, such as in the case of the wreck *Amsterdam* (Marsden 1985), the processes of wood deterioration are predominantly biological. Wood will not be degraded by wood borers due to the limited supply of dissolved oxygen within sediments thus preventing their respiration (Turner and Johnson 1971; Becker 1971). Instead deterioration of wood will be biologically mediated though the action of microorganisms, which can survive in the anoxic conditions typically found within marine sediments. Microbial deterioration in marine sediments is caused to a certain extent by soft rot fungi, which can survive in environments with limited amounts of oxygen (Blanchette et al. 1990). However, deterioration is predominantly caused by bacteria and is a very slow process; in the right circumstances archaeological wood can survive for tens of thousands of years (Björdal 2000). The predominant bacteria causing deterioration in waterlogged and anoxic environments have, as yet, not been formally identified (Helms 2008) but are termed "erosion bacteria" due to the way they enter and erode the wood cell wall leaving a distinct erosion pattern which can be identified through microscopy (Singh and Butcher 1991). Fortunately for archaeologists, erosion bacteria can only degrade the cellulose within the wood cell wall and although they may modify the lignin in the compound middle lamella, they cannot completely degrade it. Hence, the lignin rich compound middle lamella survives and its form is kept intact by the degraded parts of the cell being replaced by water.

3. *The types of materials present on the site and their state of preservation*

As has been discussed, wood, when found, may be in various states of preservation due to the effects of biological deterioration. Wood borers generally affect wood in different ways depending upon the organisms responsible. The wood boring molluscs settle onto the wood as minute larva and bore a small hole (1 mm) in the surface of the wood and then grow within the wood creating tunnels between 5 and 10 mm in diameter and sometimes in excess of 50 cm long in the wood, depending upon the age and species of organism (Turner 1966).

From the surface of the wood, and especially when diving, it is not always easy to see their presence as wood is often covered in weed, or other matter making the microscopic holes very difficult to see (Figure 1-3). The effects of the crustacea can be a little simpler to discern as they tend to degrade the surface of the wood creating small galleries (Figure 1-4). Both of these organisms will cause great loss of the physical integrity of the wood. Microorganisms, such as the aforementioned soft rot and erosion bacteria, fortunately do not affect the surface details of the wood but they degrade the wood at a cellular level removing cellulose. This results in a softening of the wood and in the case of wood from prehistoric submerged settlement sites this means the artefacts have very little strength remaining.

Thus it is important to get an overview of the actual state of preservation of the wood when considering its *in situ* preservation. Simply put, is the wood in a stable enough state to be left where it is? Are there threats of further deterioration if the wood is left *in situ*? What effects will any proposed mitigation strategies have on the wood?

In terms of assessing for the effects of wood boring organisms, a simple metal probe made of a thin rod can be used and simply pressed into the wood. Often, if shipworms are present, or have previously been active in the wood, little or no resistance is met. This is unfortunately only a qualitative assessment but will give an indication of the presence of shipworms. In order to assess the overall state of preservation of the wood that remains, density is a good parameter (Jensen and Gregory 2006). As discussed, microorganisms operate on a cellular level and, as they remove cell wall material this is replaced by water. As a result, the more

degraded the wood is, the lower the density of the wood. Density can be assessed using cores taken *in situ* with an increment borer which are subsequently processed in the laboratory. Alternately a more elaborate, yet similar method to the metal probe, can be used – the Pilodyn. The Pilodyn was originally developed to assess the extent of soft rot decay in telegraph poles in service (Hoffmeyer 1978). In terms of archaeological artefacts, Grattan et al. (1987) have described its use in assessing the condition of non-waterlogged totem

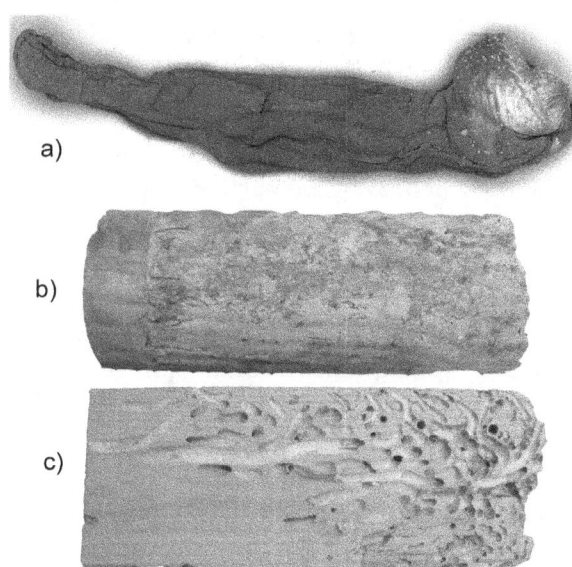

Figure 1-3 a) A typical shipworm (*Teredo spp.*) which is a wood boring bivalve mollusc. The individual was three months old and was approximately 5cm in length. The calcareous shells, which bore into the wood are to the right of the picture. b) The outside of a piece of pine exposed on a wreck site for four months. Small holes can be seen but are difficult to distinguish *in situ*. c) The inside of the same piece of pine showing the galleries formed by the shipworm (Author).

poles at the Ninstints World Heritage Site, and Clarke and Squirrell (1985) have described its use for assessing the extent of degradation of large waterlogged wooden objects.

Figure 1-4 a) A typical Gribble (*Limnoria spp*), which is a wood boring crustacean. This individual was a mature adult and was 2-3 mm long. Unlike the shipworm these attack the wood from the surface creating numerous small galleries, destroying any surface archaeological details. b) A pine wood sample after 12 months exposure showing typical attack by wood boring crustaceans (Author).

The Pilodyn (Figure 1-5) works by firing a spring-loaded blunt pin into the wood, to a maximum depth of 40 mm. The depth of penetration of the pin is indicated on a scale on the side of the instrument; the more degraded the wood, the further the pin will penetrate. The penetration reflects the shock resistance of wood, that is to say, the resistance of wood to a suddenly applied load. The energy required to overcome this resistance is a complex interaction of the various properties of wood such as wood species, density, modulus of rupture, moisture content and extent of degradation. Research (Gregory et al. 2007) has shown a good correlation between the depth of penetration of the pin and the density of waterlogged archaeological wood. In the hands of an experienced diver it is a relatively cheap, simple and robust tool for non-destructively mapping the state of preservation of timbers on the sea bed. Further to diver operated systems, remote marine geophysical techniques have also

shown the potential of these methods to not only chart the spatial distribution of wreck material *in situ*, as has been discussed, but also the state of preservation of wooden remains. Research investigating the acoustic properties of waterlogged archaeological wood showed a good correlation between the density of the wood and its acoustic impedance (Arnott et al. 2005); the principle being that the more degraded the wood, the easier it is for an acoustic source to pass through it. This method has been developed further and was trialled using a 3D Chirp sub bottom profiler on the wreck of *Grace Dieu* in the River Hamble, England (Plets et al. 2009). The results proved promising for remotely and non-destructively assessing the state of preservation of timbers.

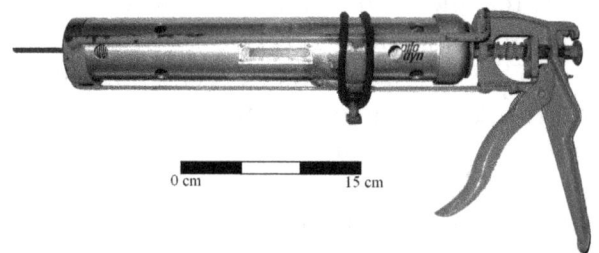

Figure 1-5 The Pilodyn wood tester (Author).

4. Strategies to mitigate deterioration and stabilise the site from natural impacts

If an initial assessment of a site's environment reveals that there are natural threats, or the site is unstable, strategies should be implemented to mitigate for these threats. It is at this stage that an overall evaluation of whether it is feasible, both practically and economically, to leave the site *in situ* should be made. It is argued that *in situ* preservation is not a panacea for managing the submerged cultural resource but just one option. Depending upon the nature of the environment and the historical and archaeological significance of a site, excavation followed either by conservation or re-deposition in a more benign environment, may be the only responsible option to ensure that it is preserved.

In terms of wooden wreck sites, the two most significant threats are the possibility of further physical deterioration, due to scour, and biological deterioration caused by wood boring organisms. Also, until we have a better understanding of the nature of the bacteria causing decay within sediments, there will always be a very slow degradation of wood due to bacterial decay.

To mitigate for these processes, sites are often covered using different methods. In the right circumstances, this can both alleviate scour and prevent the activity of wood boring organisms. In other cases, where the local environment is not conducive to simply covering, a site can be excavated and re-deposited/reburied in a more benign environment underwater or on land.

Sandbags are often used as a means of stabilising archaeological sites underwater (Gregory et al. 2008). However, their deployment is labour intensive, especially when working in areas with strong currents. Recently, maritime archaeologists and conservators have been trying to stabilise sites in situ using sediment transport to their advantage by entrapping sediment particles carried in the water column and creating an artificial seabed, or burial mound, over the threatened site. Notable examples of this are the use of artificial sea grasses on the wrecks of William Salthouse (Steyne 2009; Hosty 1988; Harvey 1996) James Matthews (Richards et al. 2009) and the Hårbølle wreck (Gregory et al. 2008). Similarly, various types of netting (shade cloth, debris netting, wind netting) have been used on several wrecks in the Netherlands (Manders 2004), Sri Lanka (Manders et al. 2004) and also trialled on the wreck James Matthews (Richards et al. 2009) and the Hårbølle wreck (Gregory et al. 2008).

The artificial sea grass and the various types of net effectively function in the same way. The plastic fronds of the artificial sea grass trap sediment particles in the water column as water passes through them. Due to friction, the water is slowed causing the sediment particles to fall out of the water column resulting in an artificial seabed/mound. In the case of netting, the net is fixed loosely over the structure to be protected, so that it billows in the water column. As with the artificial sea grass, suspended sediment in the water column passes through the net but as it does it is slowed by friction and the sand falls out of suspension and creates a mound under the net. These materials only function in the right conditions, and although it may seem common sense, the presence of sediment transport and the particle size of sediments being transported must be assessed prior to applying these methods on sites.

Should the immediate site environment not be conducive to in situ stabilisation of the site, or if, due to subsea development, a site has to be excavated, excavation and re-burial in a more benign environment is a further option. Re-burial as a means of long term storage is not a new idea and has been proposed and practised for many years around the world (De Jong 1981; Jespersen 1985, 1986; Stewart et al. 1995; Elliget and Breidhal 1991; Oxley 1998). One of the first attempts of controlled reburial of archaeological remains underwater was carried out in the 1980s. From 1980 to 1984, Parks Canada excavated the remains of the Basque whaler San Juan in Red Bay, Labrador. Following the excavation, raising and documentation of the wreck, the timbers were reburied to protect them against biological, chemical and especially physical deterioration due to ice floes (Stewart et al. 1995) What sets this early project apart from other reburial attempts of the time was that monitoring of the reburied timbers and the

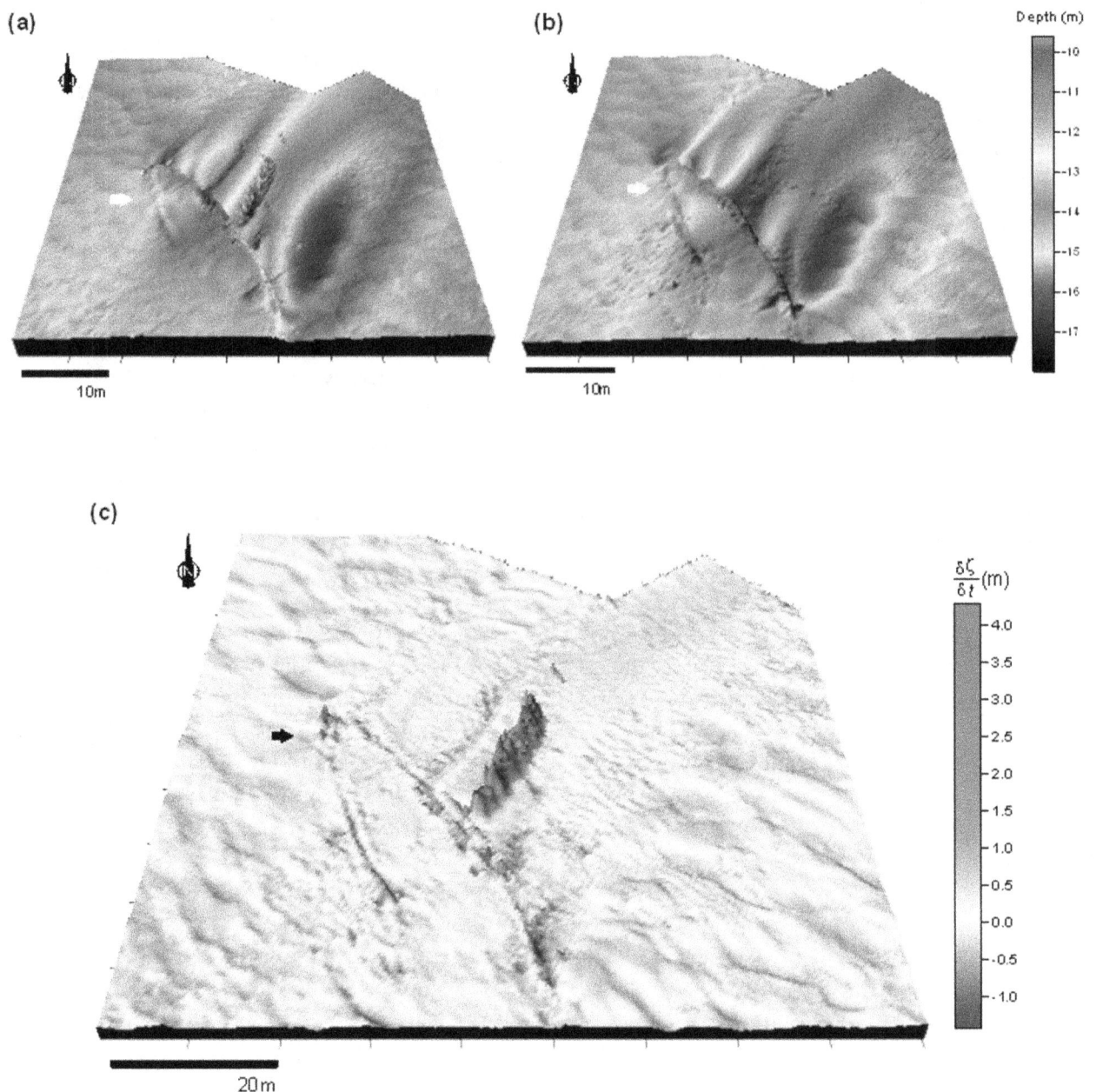

Figure 1-6 Repeat surveys over a wreck site using Multi Beam Echo Sounder. Subtracting of the data sets can show if there has been a net accumulation or erosion of sediments around a wreck or archaeological site (Rory Quinn).

surrounding reburial environment was planned from the outset. Sandbags and the ballast from the ship were used to construct an underwater cofferdam where the timbers were placed in several layers, each separated by a layer of sand. Modern wood blocks were placed alongside each layer for subsequent removal and analysis and a series of sealed dipwells installed to enable pore water samples from the mound to be removed for analysis. The burial mound was then covered with a heavy-duty plastic tarpaulin anchored by concrete filled rubber tyres. The author was fortunate to visit the site with Parks Canada in 1997 and examination of archaeological and sacrificial timbers from the burial mound showed the mitigation strategy to have been successful.

A similar project building on this work was the re-burial of artefacts from the wreck of *Fredericus* (1719) in the Swedish island port of Marstrand. Archaeological

8

investigations were initiated in the harbour because of the need to reinforce the quay. Two major investigations were undertaken; excavation of the wreck of the frigate *Fredericus*, sunk in a battle between Sweden and Denmark, and an investigation of an area alongside the quay, which revealed cultural remains dating back to the seventeenth century. These two excavations yielded approximately 10,000 artefacts. Full conservation treatment of all excavated artefacts was considered both impractical and unnecessary from an archaeological perspective and it was decided that 85-90% of the finds should be re-buried after proper archaeological documentation (Bergstrand and Nyström Godfrey 2007). Two trenches were dug for the finds in a "culturally sterile" part of the harbour. One trench was used for metal objects (mainly iron) with the other used for organic materials and silicates, with all finds being covered by at least 50cm of clay/sand sediment in 2002. The depth of burial was based on previous experiments which sought to identify the optimal burial depth for materials (Gregory 1999).

To investigate the efficacy of the reburial an umbrella research project was initiated to monitor the burial environment and to determine the effects of this environment on a range of material types to provide information which links environmental parameters and material degradation.

Four of the sub-projects are investigating the effects of the burial environment on modern materials analogous with materials commonly encountered on archaeological sites including: glass and ceramics, metals (iron and copper alloys), wood (oak, spruce and birch) and other organic materials including leather and various textiles. A fifth sub-project is investigating the stability of modern packing and labelling materials that are needed to

separate and identify archaeological material during excavation, wet storage and reburial. Since reburied artefacts are regarded as stored objects it is important to still be able to identify them after many years.

The model and packing materials were similarly covered with 50 cm of sediment from the surrounding area, except for the wood and metals samples, some of which were left exposed above the seabed as "negative" controls. The samples could then be retrieved for analysis at regular intervals to determine the impacts of the burial environment on these material types. In order to determine the long-term effects of reburial, sufficient samples have been reburied to allow sampling to continue for up to 48 years. Samples have already been recovered after 1, 2 and 3 years and it is planned that further samples will be recovered after 6, 12, 24 and 48 years. The last sub-project is examining the logistics, handling and transport of material preceding the reburial. The aim is twofold; to analyse and evaluate the logistic

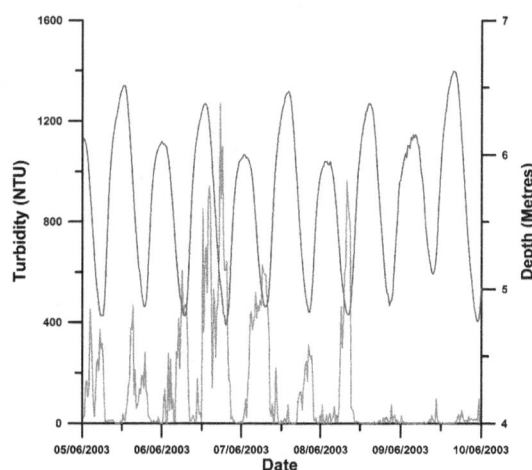

Figure 1-7 Graphs showing Turbidity measurements and depth measurements. In this instance there was a good correlation between the ebb and flood tides over the site and sediment transport, as recorded by increased turbidity (Author).

processes during the Marstrand project and to prepare a planning template that could be used for any future reburial projects.

5. Monitoring of a site and implemented mitigation strategies

In situ preservation should not stop once the site has been stabilised. Monitoring of stabilised sites is necessary to ensure continued stability. Furthermore, although a newly discovered site may be relatively stable and thus not immediately require any active mitigation strategies, environmental and/or physical changes may occur which necessitate mitigation strategies at a later date. In this context, monitoring is essential. As discussed, shipwrecks exist in a dynamic equilibrium with their environment and subsequent changes may occur through storm events or impacts of a cultural nature. This is equally valid for sites where active mitigation strategies, such as reburial, have been implemented.

As with the various processes of deterioration, monitoring should consider the two broadly different environments of open seawater and within the seabed. Within the open seawater we are concerned with physical and biological processes of deterioration namely sediment transport (erosion/accretion) and the activity of wood boring organisms. Quinn (2006) has shown the potential application of marine geophysics to monitoring the net effects of sediment transport over a wreck site. Repeat surveys carried out at different times using multi beam echo sounder, were digitally "subtracted" from each other in order to map where there were areas of net accretion and net scour of sediment (Figure 1-6). Although this shows "formation products" rather than "processes", in terms of *in situ* preservation it provides a reproducible method to quantify changes over the entirety of a site. In order to study ongoing sedimentary processes, current profilers and sediment sampling (through coring or using sediment

traps) can be placed on sites in order to model the likelihood of sediment transport (Gregory et al. 2008). The presence of actual suspended particulate matter in the water column can also be monitored using turbidity sensors/loggers (Figure 1-7). This is a relatively simple method of ascertaining if there is sediment transport and in particular when considering the use of the previously described artificial sea grass or netting materials to stabilise a site. In terms of monitoring the presence and activity of wood boring organisms over a site it is, as discussed, not always easy to monitor their activity directly on exposed timbers. However, this can simply be monitored by the placement of sacrificial blocks of modern wood around a site and recording their presence or absence. If they are present it is highly likely that any newly exposed timbers will also be colonised.

Nearly all biogeochemical processes in young (i.e. during early diagenesis) sediments are directly or indirectly

Figure 1-8 Diagram showing the various electron acceptors utilised by bacteria in typical littoral marine sediments (Author).

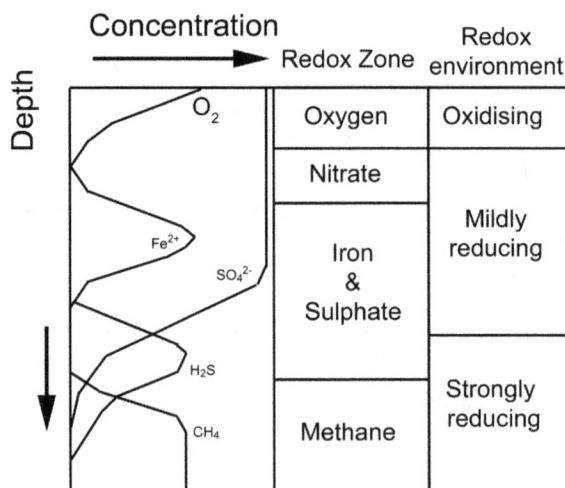

connected with the degradation of organic matter (Rullkötter 2000). Organic matter may be produced by algae and other organisms in open water, which subsequently sinks to the seabed and

10

becomes incorporated within the sediment. It may also be the remains of plant material such as eelgrass or seaweed or shipwreck material deposited within the sediment.

The utilisation of the organic matter by organisms within sediments involves oxidation-reduction (Redox) reactions (Schulz 2000). These reactions follow a well-documented succession (Figure 1-8) with various chemical species (electron acceptors) being utilised based on the amount of energy they yield (Froelich et al. 1979).

From the pool of potential electron acceptors, the microbial community selects the one that maximises energy yield from the available substrate. This is partly due to metabolic regulation within a single population and partly due to the competition between several populations with diverse metabolic capabilities. In marine sediments, the sequence of electron acceptor utilisation can be observed spatially in horizontal layers of increasing depth. In typical coastal marine sediment, only the first few millimetres of the sediment are oxygenated, though bioturbation by invertebrates and advection may extend this oxygenated zone downward. For a few centimetres under the oxygenated zone, nitrate serves as the electron acceptor followed by manganese and iron oxides. Below this, sulphate is the principal electron acceptor and sulphate reduction is often the dominant process in shallow marine sediments due to the high concentrations of sulphate in seawater. Methanogenesis is usually confined to the sulphate-depleted deeper sediment layers, though the generated methane may diffuse upward into the zone of sulphate reduction. Thus, the deterioration of organic matter still occurs in anoxic environments due to the activity of anaerobic organisms, albeit at a slower rate.

Thus in terms of monitoring within sediments, the dissolved oxygen content, concentrations of various chemical species, porosity and organic content of the sediment can all yield information about the ongoing biogeochemical processes in the sediment and the rate of deterioration of organic matter. A monitoring programme based on these parameters was trialled in Marstrand on the reburial of the *Fredericus* project and full details are given in Gregory (2007). Using a combination of data logging of pore water in dipwells and analysis of sediment cores, the environment in the burial trench at Marstrand was monitored between 2002 and 2005. Summarily, the results showed that during this period:

- The dissolved oxygen content was seen to be suboxic ($0.1 - 0.3$ mg dm^{-3}) after the first few centimetres within the cores and thereafter bordering on anoxic (<0.01 mg dm^{-3}).
- Sediments were strongly reducing with redox potentials between -160 to -250 mV (vs. SHE).
- The predominant processes ongoing in the sediments were sulphate reducing – especially at the depths where archaeological material had been reburied.
- The reburial sediments were primarily of a low porosity (0.4) sandy nature (by observation), with an organic matter content of <5%.
- From the set of core data collected in April 2006, it is apparent that there is still sulphate available for deterioration of organic matter. However, an initial diffusion model shows that the supply equates to a turnover of organic matter of 20 g of organic matter per square meter of sediment per year (2 Kg per m^2 per 100 years). This equates to the content of approximately 7 cm of sediment with 5% organic content, as has been measured in the reburial trench at Marstrand

As the results show, even though the primary deterioration processes are slow there is still ongoing deterioration of organic material. The depth of burial of 50 cm was selected based on previous experience and experiments. However, the sediment type used in Marstrand was slightly different in that there was generally less "natural" organic material in the sediments from Marstrand and that sulphate reduction was still ongoing in depths deeper than 50 cm.

Further to monitoring of the biogeochemical processes, in order to check what is happening to wooden materials, small sacrificial samples of wood should be included as part of a monitoring programme as the rate and cause of deterioration can be assessed microscopically in order to confirm biogeochemical monitoring of sediments, as was also carried out in the project (Gjelstrup et al. 2009).

Conclusion

By taking such a process based approach when assessing sites, it should be possible to identify the extent of a site above and below the sediment. Threats to a site, such as physical sediment transport and activity of wood boring organisms can be identified and where possible methods can be implemented to mitigate for these, through re-burial *in situ*, or excavation and re-deposition in a more benign environment on land or underwater. Following this, sites should be periodically monitored to assure that no significant changes have occurred to the site. Monitoring feeds back into the system, providing information on any changes and if the results reveal that the site is still under threat and it is not deemed cost effective to stabilise it *in situ,* then there is at least evidence for alternative methods to be sought. In this manner it is believed that the underwater cultural heritage may remain underwater out of sight but at least not out of mind.

Acknowledgements

The author would like to thank the organisers of the conference, and the session on *in situ* preservation, for providing the opportunity to present this work. A draft version of this paper was kindly read by Dr Rory Quinn, who also kindly provided images of his work for inclusion in this paper. The opportunity to travel to Australia and participate in the conference was made possible with the aid of funding from the Danish Directorate for Cultural Heritage (KUAS), the National Museum of Denmark, the Australasian Institute for Maritime Archaeology and Flinders University.

References

Anthony, D. and J.O. Leth
2002   Large-Scale Bedforms, Sediment Distribution and Sand Mobility in the Eastern North Sea Off the Danish West Coast, *Marine Geology* 182: 247-263.
Arnott, S.H.L., J.K. Dix, A.I. Best, and D.J. Gregory
2005   Imaging of Buried Archaeological Materials: The Reflection Properties of Archaeological Wood, *Marine Geophysical Researchers* 26:135-144.

Becker, G.

1971   On the Biology, Physiology and Ecology of Marine Wood Boring Crustaceans, In *Marine Borers, and Fouling Organisms of Wood*, E.B. Gareth Jones, S.K. Eltringham, editors, pp. 304-320, Portsmouth, England.

Bergstrand, T. and I. Nyström Godfrey

2007   *Reburial and Analyses of Archaeological Remains: Studies on the effect of reburial on archaeological materials performed in Marstrand, Sweden 2002 – 2005 The RAAR Project*, Kulturhistoriska dokumentationer Nr 20, Bohuslans Museum, Uddevalla, Sweden.

Björdal, C.G.

2000   *Waterlogged Archaeological Wood – Biodegradation and Its Implications for Cconservation*, Acta Universitatis Agriculturae Sueciae, Silvestria 142, Swedish University of AgriculturalSciences, Uppsala.

Björdal, C.G. and T. Nilsson

2009   Reburial of Shipwrecks In Marine Sediments: A Long Term Study On Wood Degradation, *Journal of Archaeological Science* 35: 862-872.

Blanchette, R.A., T. Nilsson, G. Daniel and A. Abad

1990   Biological Degradation of Wood, In *Archaeological Wood Properties, Chemistry and Preservation*, Rowell, R.M. and Barbour, R.J., editors, pp.158-161, Advances in Chemistry Series 225, Washington: American Chemical Society.

Cederlund, C. O. and F. Hocker

2006   *Vasa 1: The Archaeology of a Swedish Warship*, National Maritime Museum's of Sweden, Stockholm.

Clarke, R.W. and J.P. Squirrell

1985   The Pilodyn: An Instrument for Assessing the Condition of Waterlogged Wooden Objects, *Studies in Conservation* 30: 177-183.

De Jong, J.

1981   The Deterioration of Waterlogged Wood and Its Protection in Soil. In *Conservation of waterlogged wood: International symposium on the conservation of large objects of waterlogged wood*, De Vries-Zuiderbaan, editor, pp. 57-68. Netherlands.

Elliget, M. and H. Breidhal

1991   *A Guide to the Wreck of the Barque William Salthouse*. Melbourne.

Froelich, P.N., G.P. Klinkhammer, M.L. Bender, N.A. Luedtke, G.R. Heath, D. Cullen, P. Dauphin, D. Hammond, B. Hartman and V. Maynard

1979   Early Oxidation of Organic Matter in Pelagic Sediments of the Eastern Equatorial Atlantic: Suboxic Diagenesis, *Geochim. Cosmochim. Acta* 43: 1075-1090.

Grattan, D.W., W. Bokman and C.M. Cook

1987   Scientific Examination of Totem Poles at Ninstints World Heritage Site, *Journal of IIC Canadian Group* 12: 43-57.

Gregory, D.

2007   Environmental Monitoring. In *Reburial and Analyses of Archaeological Remains. Studies on the effects of reburial on archaeological materials performed in Marstrand, Sweden 2002-2005, The RAAR project*, T. Bergstrand and I. Nyström Godfrey, editors, pp. 59-90, Kulturhistoriska dokumentationer Nr 20, Bohuslans Museum, Uddevalla, Sweden.

Gregory, D., P. Jensen, H. Matthiesen and K. Strætkvern
2007   The Correlation Between Bulk Density and Shock Resistance of Waterlogged Archaeological Wood Using the Pilodyn, *Studies in Conservation* 52: 289-298.

Gregory, D., R. Ringgaard and J. Dencker
2008   From a Grain of Sand a Mountain Appears. Sediment Transport and Entrapment to Facilitate the *In Situ* Stabilisation of Exposed Wreck Sites, *Maritime Newsletter from Denmark, Syddansk Universitet* 23: 15-23.

Gregory, D.J.
1999   Re-Burial of Timbers In the Marine Environment as a Means of Their Long-Term Storage: Experimental Studies In Lynæs Sands, Denmark, In *The International Journal of Nautical Archaeology* 27(4):343-358.

Grenier, R.
2006   Introduction: Mankind, and at Times Nature, are the True Risks to Underwater Cultural Heritage. In *Heritage at Risk –Special Edition- Underwater Cultural Heritage at Risk,* R. Grenier, D. Nutley & I. Cochran, editors, pp. 58-60, International Council on Monuments and Sites, Paris, France.

Harvey, P.
1996   A Review of Stabilisation Works on the Wreck of the *William Salthouse* in Port Phillip Bay, *Bulletin of the Australasian Institute for Maritime Archaeology* 20(2):1-8.

Helms, A.C.
2008   *Bacterial Diversity in Waterlogged Archaeological Wood,* Unpublished PhD. Thesis. The Bio Centre, Danish Technical University, Kongens Lyngby, Denmark.

Hoffmeyer, P.
1978   The Pilodyn Instrument As a Non-Destructive Tester of the Shock Resistance of Wood, In *Proceedings of the Fourth Symposium on Non Destructive Testing of Wood,* Vancouver, W.A. Pullman, editor, pp. 47-99, Washington DC.

Hosty, K.
1988   Bagging the *William Salthouse*: Site Stabilization Work on the *William Salthouse, Bulletin of the Australasian Institute for Maritime Archaeology* 12(2): 13-16.

Ward, I.A.K., P. Larcombe and P. Veth
1999   A New Process-Based Model for Wreck Site Formation, *Journal of Archaeological Science* 26: 561-570.

Jensen, P. and D.J. Gregory
2006   Selected Physical Parameters to Characterize the State of Preservation of Waterlogged Archaeological Wood: A Practical Guide for Their Determination, *Journal of Archaeological Science* 33: 551–559.

Jespersen, K.
1985   Extended Storage of Waterlogged Wood in Nature. In *Proceedings of the ICOM Waterlogged Wood Working Group Conference,* D.W. Grattan and J.C. McCawley, editors, pp. 39-54, Ottawa.

Jespersen, K.
1986   Extended Storage of Waterlogged Wood, When Excavated and *In Situ.* In *Preventive Measures During Excavation and Site Protection,* M. McCarthy, editor, pp.113-131, Rome.

Manders, M.R., W.M. Chandraratne, A.M.A. Dayananda, R. Muthucumarana, K.B.C. Weerasena and K.D.P. Weerasingha
2004   The Physical Protection of a 17th century VOC Shipwreck in Sri Lanka, *Current Science* 86(9): 101-107.

Manders, M.R.
2004   Protecting Common Maritime Heritage. The Netherlands Involved In Two EU-Projects: MoSS and BACPOLES, In *Mediterraneum* Vol.4. Protection and Appraisal of Underwater Cultural Heritages, F. Maniscalco, editor, pp.279-292.

Marsden, P.
1985   *The Wreck of the Amsterdam*, Second edition, Hutchinson, 69-70, London.

Oxley, I.
1998   The *In Situ* Preservation of Underwater Sites. In *Preserving Archaeological Remains* In Situ, M. Corfield, P. Hinton, T. Nixon and M. Pollard, editors, pp. 159-173. London.

Plets, R.M.K., J.K. Dix, J.R. Adams, J. M. Bull, T.J. Henstock, M. Gutowski and A.I. Best
2009   The Use of High-Resolution 3D Chirp Sub-Bottom Profiler for the Reconstruction of the Shallow Water Archaeological Site of the *Grace Dieu* (1439), River Hamble, *Journal of Archaeological Science* 36: 408-418.

Quinn, R., J.R. Adams, J.K. Dix and J.M. Bull
1998   The *Invincible* (1758) Site - An Integrated Geophysical Assessment, *International Journal of Nautical Archaeology* 27(3): 126-138

Quinn, R., J.M. Bull, J.K. Dix and J.R. Adams
1997   The *Mary Rose* Site - Geophysical Evidence for Palaeo-Scour Marks, *International Journal of Nautical Archaeology* 26(1): 3-16.

Quinn, R.
2006   The Role of Scour in Shipwreck Site Formation Processes and the Preservation of Wreck-Associated Scour Signatures in the Sedimentary Record, *Journal of Archaeological Science* 33(10): 1419-1432.

Richards, V., I. Godfrey, R. Blanchette, B. Held, D. Gregory and E. Reed
2009   *In situ* Monitoring and Stabilisation of the *James Matthews* Shipwreck. In *Proceedings of the 10th ICOM Group on Wet Organic Archaeological Materials Conference*, K. Strætkvern and D.J. Huisman, editors, pp. 1130160, Nederlandse Archaeologische Rapporten 37, Amersfoort, The Netherlands.

Richards, V., I. MacLeod and P. Morrison
2009   Corrosion Monitoring and Environmental Impact of Decommissioned Naval Vessels As Artificial Reefs. In In Situ *Conservation of Cultural Heritage : Public, Professionals and Preservation. Archaeology from Below: Engaging the Public.* V. Richards and J. McKinnon, editors, pp.50-67, Past Foundation.

Rullkötter, J.
2000   Organic Matter: The Driving Force for Early Diagenesis. In *Marine Geochemistry* Schulz HD & Zabel M, editors, pp.129-153, Springer-Verlag, Berlin.

Schulz, H.D.
2000   Redox Measurements in Marine Sediments, In *Redox: Fundamentals, Processes and Applications*, Schüring J, Schulz HD, Fischer WR, Böttcher J and Duijnisveld WHN, editors, pp. 235-246, Springer Verlag, Berlin.

Singh, A.P. and J.A. Butcher
1991   Bacterial Degradation of Wood Cell Walls: A Review of Degradation Patterns, *Journal of the Institute of Wood Science* 12:143

Stewart, J., L.D. Murdock and P. Waddell
1995   Reburial of the Red Bay Wreck as a Form of Preservation and Protection of the Historic Resource. In *Materials issues in art and archaeology IV*, Materials research society symposium proceedings, vol. 352. Mexico 1994.

Steyne, H.
2009   Cegrass™, Sand & Marine Habitats: A Sustainable Future for the *William Salthouse.* In In Situ *Conservation of Cultural Heritage: Public, Professionals and Preservation*,V. Richards and J. McKinnon, editors, pp.40-49, Past Foundation.

The Insitute of Field Archaeologists, United Kingom. Maritime Affairs Group Bulletin.
2008   http://ifamag.wordpress.com/mag-bulletins/

Turner, R.D. and A.C. Johnson
1971   Biology of Marine Wood Boring Molluscs. In *Marine Borers, and Fouling Organisms of Wood*, E.B. Gareth Jones, S.K. Eltringham, editors, pp. 259-296, Portsmouth, England.

Turner, R.D.
1966   *A Survey and Illustrated Catalogue of the Teredinidae*, Harvard University, Cambridge, Mass.

2 Developing Methodology for Understanding In Situ Preservation and Storage from a Practitioner Perspective

Nicole Ortmann

Flinders University, Department of Archaeology, Program in Maritime Archaeology, GPO Box 2100, Adelaide, South Australia, 5001

In situ forms of preservation and storage are often emphasised as the preferred option under most circumstances for conserving underwater and waterlogged cultural heritage for future generations. With the adoption of the UNESCO 2001 Convention, it is imperative that current approaches and trends among practitioners be identified and evaluated. Key to understanding these ideas is developing a methodology that will provide a valid analysis. By combining a literature review and a practitioner questionnaire, current concepts and beliefs about *in situ* preservation and storage should be highlighted, resulting in a solid base for further discussions about the safeguarding of underwater cultural heritage.

Introduction

'[H]istory is, in large part, a catalogue of examples of in situ [sic] preservation' (Holden et al. 2006:59)

As the quote above attests, *in situ* preservation occurs naturally, to an extent, within archaeology as a whole. This preservation is often even more pronounced in waterlogged and submerged environments. From the discovery of Swiss Lake villages in the 1850s (Desor 1865; Delgado 1997:233-235) and Roman shipwrecks in the early twentieth century (Ucelli 1950; Delgado 1997:233; Muckelroy 1998:29-31) to the modern and high-tech explorations of the deep ocean that have revealed ancient wrecks such as the Skerki vessels (McCann 2001:257; McCann and Oleson 2004), archaeology has been aware of the innate natural ability of water, especially waterlogged sediments, to preserve a wide range of cultural materials. In the last few years, trends in submerged cultural heritage management have been towards *in situ* preservation and storage for a number of reasons, such as financial and curatorial considerations (Stewart, Murdock and Waddell 1995:793; Corfield 1996:33; Oxley 1998:159). But while archaeological sites in submerged and wetland areas

continue to be discovered, proving that natural preservation is possible, the chemical, biological and physical mechanisms behind these discoveries have only recently begun to be explored (Corfield 1996:32; Caple, Dungworth and Clogg 1997:57; Oxley 1998:159; Manders 2004:279).

Textbooks focusing on underwater archaeology or heritage management often include sections about *in situ* preservation (The Nautical Archaeology Society 1992:332; Babits and Van Tilburg 1998:590; Green 2003:470). A review of these texts shows that the concept is poorly defined. This can lead to an understanding by newcomers to the archaeological field that wet sites reach equilibrium with their environment and therefore, if not physically disturbed, will remain stable over time. While to an extent this is true and may be preferable to any type of disturbance, the "don't touch" attitude does not necessarily constitute *in situ* preservation. The site will change as the environment around it changes and actions must be taken, either through active intervention or monitoring, to confirm that these changes are not affecting preservation.

Recently, *in situ* forms of preservation and storage have been consistently emphasised as the preferred option under most circumstances for preserving submerged and waterlogged cultural heritage for future generations (The Nautical Archaeology Society 1992:332; International Council on Monuments and Sites 1996:1-5; Babits and Van Tilburg 1998:590; United Nations Educational, Scientific and Cultural Organization 2001:56-61; Green 2003:470; Bergstrand and Nyström Godfrey 2007:7 & 15). The United Nations Educational, Scientific and Cultural Organization (UNESCO) underscores the use of *in situ* methods in its 2001 Convention on the Protection of the Underwater Cultural Heritage (United Nations Educational, Scientific and Cultural Organization 2001: 5 & 58-60) as does the 1996 Charter for the Protection and Management of the Underwater Cultural Heritage adopted by the International Council on Monuments and Sites (International Council on Monuments and Sites 1996:2). Many other organisations, while not formally installing *in situ* preservation into their by-laws or constitutions, still stress the importance of this concept in their educational programmes; the Nautical Archaeology Society (NAS) in the United Kingdom is one such group. If *in situ* methods are to be promoted as the primary means of preserving underwater cultural heritage, they must be shown to be based on sound scientific premises, or it could be difficult to argue that *in situ* preservation and storage methods are truly in the best interest of the artefacts, features and sites.

In situ preservation is based on the concept that certain environments naturally produce situations capable of slowing deterioration. As Holden et al. (2006:59) indicate, it is this very process that allows archaeologists to uncover the past through excavation. Early reburial schemes, such as those used at Red Bay, stem in part from this idea (Stewart, Murdock and Waddell 1995:794). During the last decade, studies have been undertaken to explore the idea of *in situ* preservation and storage in order to test the assumptions made about the preservative nature of sediment coverage in waterlogged environments, either natural or through reburial. In Australia, work on several wrecks has been particularly important in experimenting with the use of sacrificial anodes on metals and reburial schemes (McCarthy 1987; Hosty 1988; Nash 1991; MacLeod 1993; Moran 1997; MacLeod 1998; Godfrey et al. 2005; Winton and Richards 2005). For example, the use of sacrificial anodes and cathodic protection for iron artefacts by MacLeod on the *Sirius* site (MacLeod 1996:111) was originally intended to provide increased stability in conjunction with conventional forms of retrieval and treatment. Since then, anodes have been used to protect sites *in situ* where there has been no intention of retrieving remaining materials (MacLeod 1998:81; MacLeod 1995:53; Heldtberg, MacLeod and Richards 2005:75; MacLeod et al. 2005:53). The inclusion of conservation scientists with various chemical and biological backgrounds led in some instances to the adoption of well-established scientific principles into conservation strategies. Programmes such as Reburial and Analyses of Archaeological Remains (RAAR) in Marstrand, Sweden (Bergstrand and Nyström Godfrey 2007) and Monitoring, Safeguarding and Visualizing North-European Shipwreck Sites (MoSS), involving a number of European nations such as Finland, Denmark, the Netherlands, Germany, Sweden and the United Kingdom (Cederlund 2004), have been carried out in the field on a number of sites. Laboratory studies by Björdal, Nilsson and others in Sweden have also shown that there is merit in pursuing *in situ* methods (Björdal and Nilsson 1999; Björdal, Nilsson and Daniel

1999; Björdal, Daniel and Nilsson 2000; Björdal and Nilsson 2002; Björdal, Nilsson and Petterson 2007; Björdal and Nilsson 2008).

As more emphasis is placed on protecting submerged cultural heritage, it becomes crucial to understand how *in situ* preservation and storage is perceived and utilised to protect these resources. With UNESCO having taken effect on 2 January 2009 and impacting 26 signatory states, the prevalence of *in situ* programmes stands to increase and the methods used could impact protection of submerged cultural heritage in many ways. In light of these global developments, what is known about *in situ* preservation and storage? What projects currently use these methods and how successful are they? What contributions are being made by other disciplines and how are they incorporated into the fields of archaeology and cultural resource management?

It is not just the archaeology and the science, however, which needs to be addressed. Central to implementing *in situ* preservation and storage is an understanding of current attitudes towards, and uses of, *in situ* preservation methods among practitioners in light of recent research. Who is using *in situ* preservation methods and why? If practitioners are, as a whole or in part, not using *in situ* preservation and storage, what are the reasons? What forms are the most prevalent and why?

Given the two-fold nature of this research, mechanisms to address these varied questions are required. This paper explores the methodology developed to understand the current research and methods of *in situ* preservation and storage alongside the attitudes and perspectives of the practitioners. It is hoped that by combining what is known about *in situ* techniques with ways of thinking about *in situ* preservation and storage, trends crucial to building the future of the practice will be identified.

Defining the Field

In order to explore this topic in more depth, it was necessary to create definitions that would allow the subject matter to be limited to a manageable size. Waterlogged sediment occurs in a variety of different areas, including urban and rural terrestrial sites. In the United Kingdom, for example, English Heritage has funded substantial work on wetland sites dating from the Neolithic through the Medieval period (Corfield 1996; Goodburn-Brown and Hughes 1996; Caple, Dungworth and Clogg 1997; Caple 1998). As the intent of this research was to explore underwater and maritime heritage, sites such as these will not be addressed in this study. To this end, the following three definitions were applied to determine the types of sites to be explored throughout this research.

Maritime Archaeology

The study of human interaction with the sea through seafaring. This includes not only the vessels themselves, but port and harbour structures; fishing, whaling and other maritime subsistence activities; lighthouse and shore-based structures that aid in seafaring; and any other type of site that has connections to the use of the sea and its resources by humans.

Underwater Site

Any site, feature or artefact found in a body of water, whether it be a lake, river or sea; these sites may include those which have become inundated over time and are currently underwater, such as habitation or ceremonial sites.

Waterlogged Terrestrial Site

Any site that may now be treated as a terrestrial site, but was at some previous time under any body of water such as a lake, river or sea and which people interacted with as a water body for the purposes of transport, subsistence, economy or ceremony. These sites will not include sites which have always been terrestrial but yet waterlogged unless they can be clearly related to the maritime landscape through the above definition of maritime archaeology.

It was also useful to define *in situ* preservation and storage. Submerged and buried maritime heritage exists in an environment that, without disturbance, is conducive to long-term preservation of a variety of archaeological materials (Corfield 1996:32; Bergstrand and Nyström Godfrey 2007:10). Once these sites are disturbed, chemical, biological and physical forces begin to destroy the fragile stability. *In situ* preservation and storage aims at restoring this stability by slowing down the mechanisms of deterioration and degradation (Corfield 1996:32). It is important to note that these techniques do not stop deterioration of archaeological materials. There are many ways to provide stability and often the delineation between them is blurred. The following definitions aimed at providing some clarity.

In situ *Preservation*

Any steps taken on or intervention with a site in order to extend its longevity while maintaining original context and spatial position; while artefacts and features may have been excavated and/or removed, the site itself remains in place and retains all or a majority of its original context.

In situ *Storage*

Any steps taken to preserve the physical, historical and aesthetic integrity of artefacts and features excavated from a site through the creation of a separate space where items are stored within the confines of an environment similar or deemed to be more beneficial to that from which they were removed.

Developing the Methodology

Once the field of research was characterised through the definitions, a suitable approach was needed in order to investigate current practitioners' attitudes towards *in situ* preservation and storage in a cohesive manner. Triangulation Design was chosen, specifically a convergence model that allows the qualitative and quantitative to be collected and analysed separately and the results of each to be combined to produce an interpretation (Creswell and Plano Clark 2007:64-65). The intent of this model was to provide a legitimate set of conclusions demonstrating how *in situ* preservation and storage is understood and used. The two methods used in this model were a literature review and a questionnaire. To determine what types of influences would be acting on practitioners, it was essential to understand the body of literature produced. A mixed-methods research design allowed for the literature to inform a mainly quantitative approach to collecting information about attitudes through a questionnaire. Central to this decision was the issue of interdependence: the question of practitioners' attitudes was dependent on a qualitative reading of the literature (Creswell and Plano Clark 2007:34).

A literature review is a 'systematic, explicit, and reproducible method for identifying, evaluating, and synthesizing the existing body of completed and recorded work produced by researchers, scholars, and practitioners' (Fink 2005:3). Literature reviews describe a current body of research with the objective of explaining and guiding professional practices, by identifying and developing new avenues of research or through interpreting existing literature (Fink 2005:8-10). In this research, the literature review was approached from a qualitative standpoint as the intent of the review was inductive and exploratory. Understanding what developments and ideologies may have had a hand in informing possible attitudes also proved helpful for designing the questionnaire (Creswell 2003:88-89). The following provides a description of the methodology used to locate and review the literature explored.

By applying the developed definitions of maritime, underwater and waterlogged sites, a number of databases, such as ScienceDirect, Wiley Interscience and Web of Knowledge, were explored in order to identify possible items for review. Content deemed useful was read and summarised; of particular importance were the bibliographies as these were able to generate items of interest that had not been found during initial research. These were then divided into two separate groups: those approaching *in situ* preservation and storage from a primarily archaeological stance, and those approaching the subject in terms of the chemical, physical and biological. The former category consisted mainly of archaeological surveys and excavations carried out by archaeologists based on a variety of understandings about *in situ* methods. Some projects may have had conservators present who informed the preservation process while others did not. The main delineator was that the intent of the project was to preserve remains *in situ*, regardless of what form was used. In the second category, the projects were experimental in nature, with the intent being to understand the processes behind *in situ* preservation. These studies were less focused on preserving particular underwater cultural heritage and more on understanding or developing methods and techniques.

Using an exploratory approach (Hart 1998:47), a broad understanding of both types of literature was developed that defined the history of the topic as well as the current state. Questions asked were based on those defined by Fink (2005:53). Was the research design valid and were the sources on which it was based consistent and applicable? Were the methods used appropriate and are the results yielded significant and practical? Was there an understanding of the strengths and weaknesses of the research?

Also of importance was assessing whether or not the current body of literature was accessible to the varied audience using it to inform subsequent projects and research. Here, the explanatory approach was used to explore the event, in this case, current practitioners' attitudes. What types of attitudes were prevalent within the literature and could these views be linked to the current state of research? This, along with a summative evaluation of the literature, provided a basic overview of existing research and projects. From this knowledge base, abstract ideas about *in situ* preservation and storage were reformed into questions that fashioned the questionnaire.

Certain limitations became evident during research. The amount of literature to be reviewed meant that studies deemed by the researcher to be more inclusive or significant received more attention. This introduced a particular bias into the research. The interdisciplinary nature of the literature also proved restrictive. This in itself highlighted the varied nature of the subject, which will no doubt continue to be an area for development. The language barrier also limited the extent of the review. Only literature available in either English or French was explored.

Following from the literature review, a questionnaire was developed to evaluate attitudes towards *in situ* preservation and storage by practitioners throughout the world. A number of texts were consulted (Foddy 1988; Foddy 1993; De Vaus 2002; Alreck and Settle 2004). The sampling method chosen to define participants followed the method of non-probability purposive sampling (De Vaus 2002:91). This was based on the notion that the questions to be asked require a certain amount of insider knowledge in the field of maritime and underwater archaeology, as well as conventional and *in situ* methods of preservation. A list of those invited to participate was drawn in part from the review of the literature. This provided a solid basis from which to expand as it was composed of current practitioners in the scientific, archaeological and heritage conservation and management communities. In addition to this, discussion with individuals in these communities known to the researcher identified other participants through professional relationships. The Nautical Archaeology Society in the United Kingdom agreed to circulate a request for participation to the membership and the online Museum of Underwater Archaeology (MUA) managed out of Rhode Island was amenable to posting a notice on its website. Other groups approached included members of The Conservation Digest and Sub-Arch list serves, as well as the Society for Historical Archaeology (SHA), the Australasian Institute for Maritime Archaeology (AIMA) and the American Institute for Conservation (AIC). Through this method of networking, a representative sample of practitioners covering most professions, such as research scientists, archaeologists, conservators and academics in related fields was created.

The people involved in the practice of *in situ* preservation and storage are, as previously noted, a varied group with a diverse knowledge base. They work in countries around the world in a number of different areas, such as government heritage agencies, public and private museums, university departments, not-for-profit agencies and consulting firms. As a result, the questionnaire had to be developed in such a way as to be understood by this group, not just in terms of possible language barriers, but also in terms of inclusive definitions.

Preliminary research identified two main theoretical areas to pursue in terms of survey design: attitudes and behaviours. Surveys designed to assess attitudes investigate how existing knowledge affects actions (Alreck and Settle 2004:13-14). This was intended to highlight the familiarity of practitioners with the literature about *in situ* preservation and storage and its influence on their actions. The second, behavioural survey, was intended to assess questions such as 'what, where, when and how often' (Alreck and Settle 2004:20) in order to understand the types of *in situ* preservation and storage methods previously used, those being currently used and what techniques may be used in future. It also allowed for the ability to identify changes in patterns and routines (Alreck and Settle 2004:20-21).

The developed definitions were a focal point for developing the questionnaire by focusing respondents, hopefully reducing the numbers of varied interpretations that can occur with self-administered questionnaires (De Vaus 2002:49). Defining *in situ* preservation and storage early on in the process also aided in creating indicators that would later be developed into the questions posed. Also, the questionnaire could later be analysed alongside the literature review in a consistent manner.

De Vaus (2002:45) mentions three ways of developing indicators. The first explores existing research for previously developed measures and concepts. This method was partially applicable in this instance. The literature review provided a number of definitions and concepts such as what types of *in situ* methods are being used and investigated. From this, the methods were developed into indicators of behaviour. In terms of developed measures, no research into attitudes and behaviours of practitioners in this area has been previously conducted. It was not surprising that no such measures existed.

The second method collects data in a less structured form in order to facilitate the understanding of the group to be studied. The group's thought patterns and actions can then help create questions that are relevant to the subject being explored (De Vaus 2002:45). This combined well with the third method that requires using information provided by those within the group. In this case, personal experience coupled with conversations with a number of individuals practicing both conventional and *in situ* methods helped to guide question formation. In particular, Vicki Richards, a Research Officer/Conservation Scientist with the Western Australian Museum, was able to provide many useful suggestions and criticisms.

Once the concepts and indicators were developed, the next step was to create and group questions in a way that was both logical and easily understood. Questions were formulated in order to fulfil the following: ease of comprehension by participants through clear definition of the subject (Foddy 1993:25); tolerable length and time frame for participants (De Vaus 2002:112); ease of access for participants in terms of language and delivery (Alreck and Settle 2004:183-186) and straight-forward flow of questions (De Vaus 2002:110). Originally a three-part questionnaire was conceived. This focused on the general properties of the site, the practice of *in situ*

preservation and storage and the use of monitoring. This structure stemmed from the fact that it was necessary to understand *in situ* preservation and storage from a site-specific background. As it was possible that one of the measures of use was to be based on the site itself, it was practical to define sites in terms of physical and environmental parameters as well as the types of cultural heritage that may be found within those sites.

The largest portion of the survey focused on what types of *in situ* preservation and storage are utilised and how many practitioners are using these procedures. The questions were developed on the basis that there should exist three types of practitioners: those who use *in situ* preservation, those who have used it in the past but have changed their minds and those who have never used it. It also stood to reason that there would be a number of factors influencing decisions made against the use of *in situ* preservation and that in many cases it would be unlikely to be a single reason. It was from discussions of this issue with Vicki Richards (2008, pers. comm.) that a format evolved that allowed for the use of mainly multiple response questions, including the ability for respondents to provide an answer not specified in the questionnaire. These answers could introduce novel ideas about the topic beyond the scope of closed questions.

The third section of the questionnaire focused on monitoring sites. Early on, the decision was made to allow respondents who do not use *in situ* preservation and storage to answer questions in this section. This was based on personal experience with not-for-profit volunteer organisations, such as the Underwater Archaeological Society of British Columbia and the Nautical Archaeology Society. These groups are an integral part of protecting underwater cultural heritage. Monitoring programmes are of primary importance where society mandates focus on preserving heritage. While monitoring in itself may not actively preserve the site, it is an integral part of the *in situ* process and due to the assumed cost efficiency inherent in engaging the existing volunteer base, it was assumed to be one of the more well-utilised methods.

It was after these sections were formulated that the idea for exploring the respondents' backgrounds began to germinate. As the development and use of *in situ* preservation and storage is a multidisciplinary one, patterns could possibly be brought to light about how the different careers and sectors viewed these methods. Two introductory questions were created that focused on how the respondents viewed themselves in terms of their profession and sector. Respondents would be asked to choose one answer for each question that they felt best described their situation, but were still allowed to provide their own answer. The answers to these two questions have the capacity to allow for more specific analyses to be completed.

Limitations are expected with surveys. Those identified as issues in this questionnaire include failure of participants to respond (Alreck and Settle 2004:37 & 205), individual participants interpreting questions in different ways (Foddy 1993:189), issues between the relationship of what respondents reported they did and what they actually did (Foddy 1993:3) and misapplying statistical methods to the data (Alreck and Settle 2004:269). Some of these, such as the use of the correct statistical methods, have been addressed through research and questionnaire design as well as understanding the types of questions asked. Others, such as response rates, were

Figure 2-1 First question of practitioners' questionnaire (SurveyMonkey™).

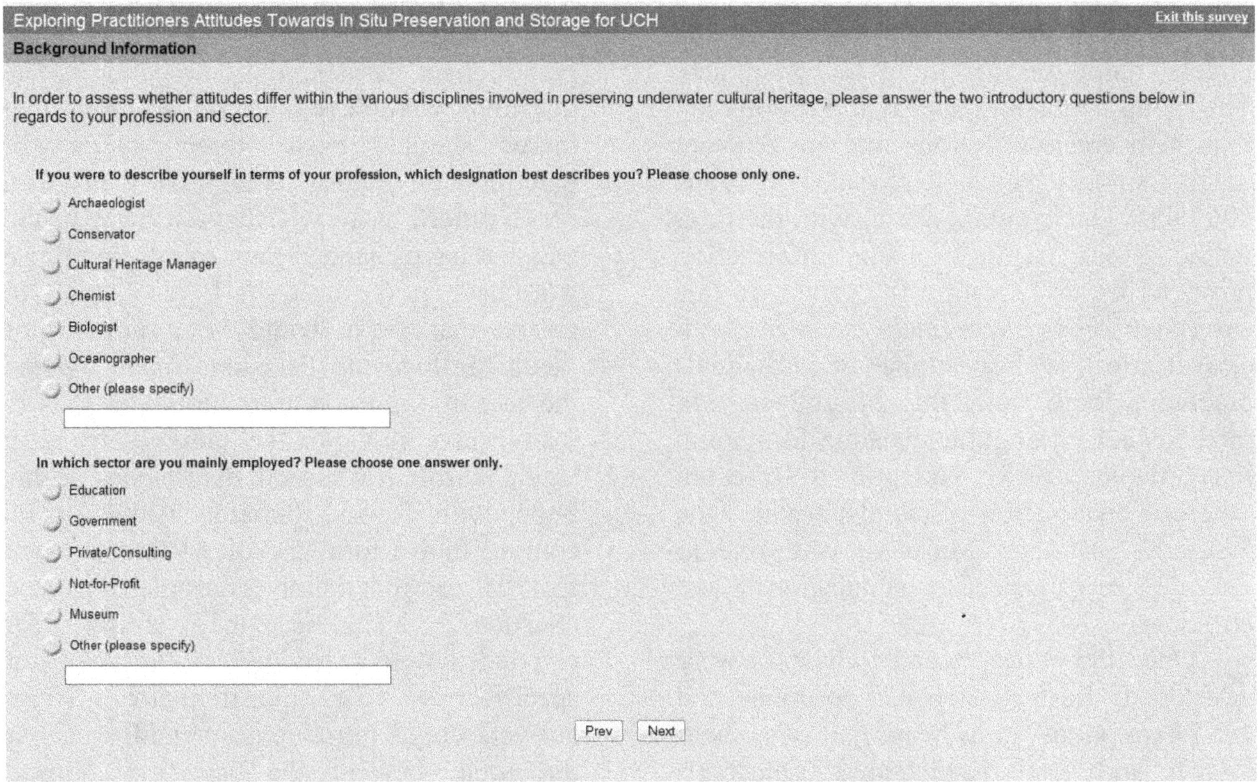
Exploring Practitioners Attitudes Towards In Situ Preservation and Storage for UCH

Exit this survey

Background Information

In order to assess whether attitudes differ within the various disciplines involved in preserving underwater cultural heritage, please answer the two introductory questions below in regards to your profession and sector.

If you were to describe yourself in terms of your profession, which designation best describes you? Please choose only one.

- Archaeologist
- Conservator
- Cultural Heritage Manager
- Chemist
- Biologist
- Oceanographer
- Other (please specify)

In which sector are you mainly employed? Please choose one answer only.

- Education
- Government
- Private/Consulting
- Not-for-Profit
- Museum
- Other (please specify)

Prev Next

accepted as inherent risks to survey methodology.

It was determined that the best delivery system for this questionnaire would be an online method, using SurveyMonkey™. This would allow participants to access the questionnaire easily and eliminate the inherent problems associated with completing paper questionnaires and return post (Alreck and Settle 2004:183). The online questionnaire also stood to be less expensive. As this would be an international questionnaire, the postage costs could have become prohibitive. This would affect the number of surveys sent, which would, in turn, affect the number received. The survey software would also allow for hard copies to

can respond as soon as they receive their link. The email system in SurveyMonkey™ permits reminders to be sent to those who haven't responded. Data is also available for analysis immediately after the questionnaire has been received as complete by the system.

SurveyMonkey™ has the added advantage of aesthetics (Figure 2-1). The questionnaire appears professionally created. SurveyMonkey™ allows for a variety of question types which is important as no one site is identical to another and it was necessary to allow people to answer in as large a range as possible. Very few questions allow only one answer; most allow the respondent to choose as many as applied

Figure 2-2 Bar graph generated from responses (SurveyMonkey™).

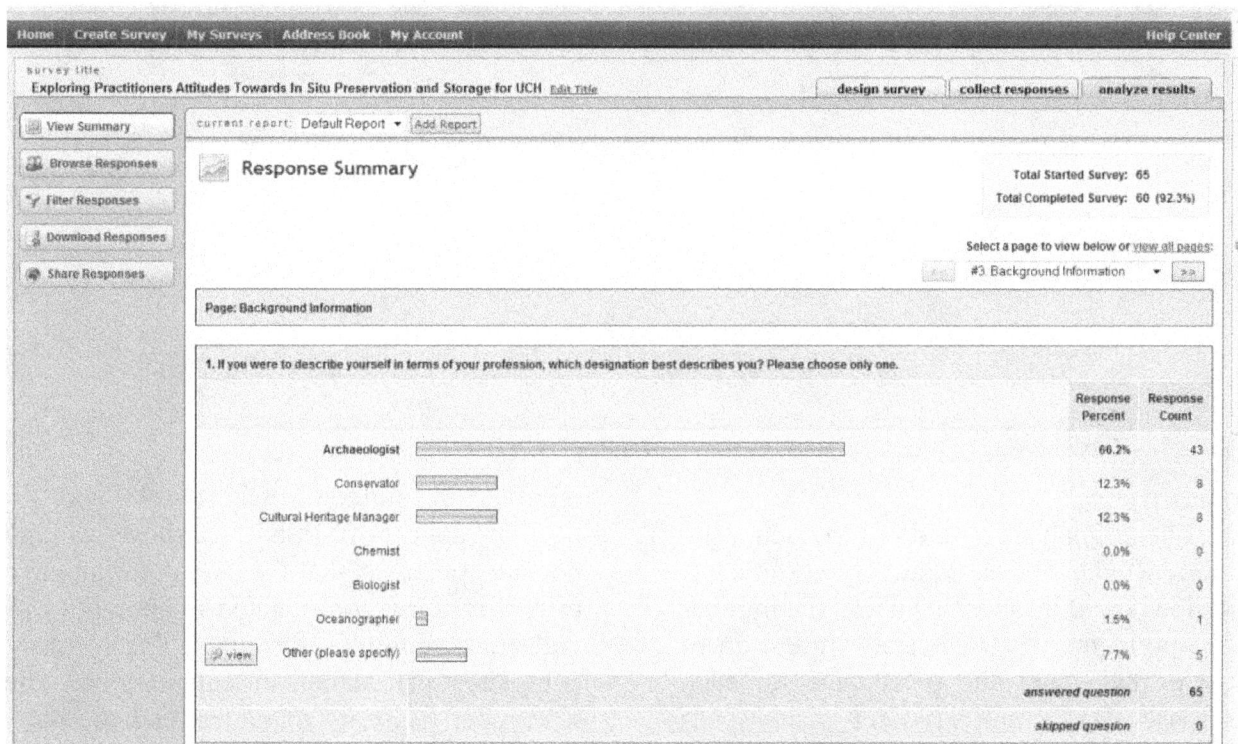

be printed for mailing.

The online delivery system allows surveys to be sent and received more quickly than relying on postal systems. Emails containing survey links are delivered five minutes after activation. Respondents

to their situation. Comment boxes are also included, which allow respondents to clarify or expand their answers. Rather than having to provide detailed instructions on a question-by-question basis, SurveyMonkey™ can apply logic formulas which guide the respondent to the next logical page depending on the answer provided. The downside to this is the time spent inputting formulas and testing the questionnaire each time a change is made. Furthermore, a second survey has to be created with question-by-question instructions for those who require paper

Package for Social Sciences), will be utilised. This requires the data to be downloaded from SurveyMonkey™ into a comma delineated worksheet. From there the data will be edited into a format accepted by SPSS™. This purports to be a time-consuming exercise, requiring all questions answered with worded responses to be

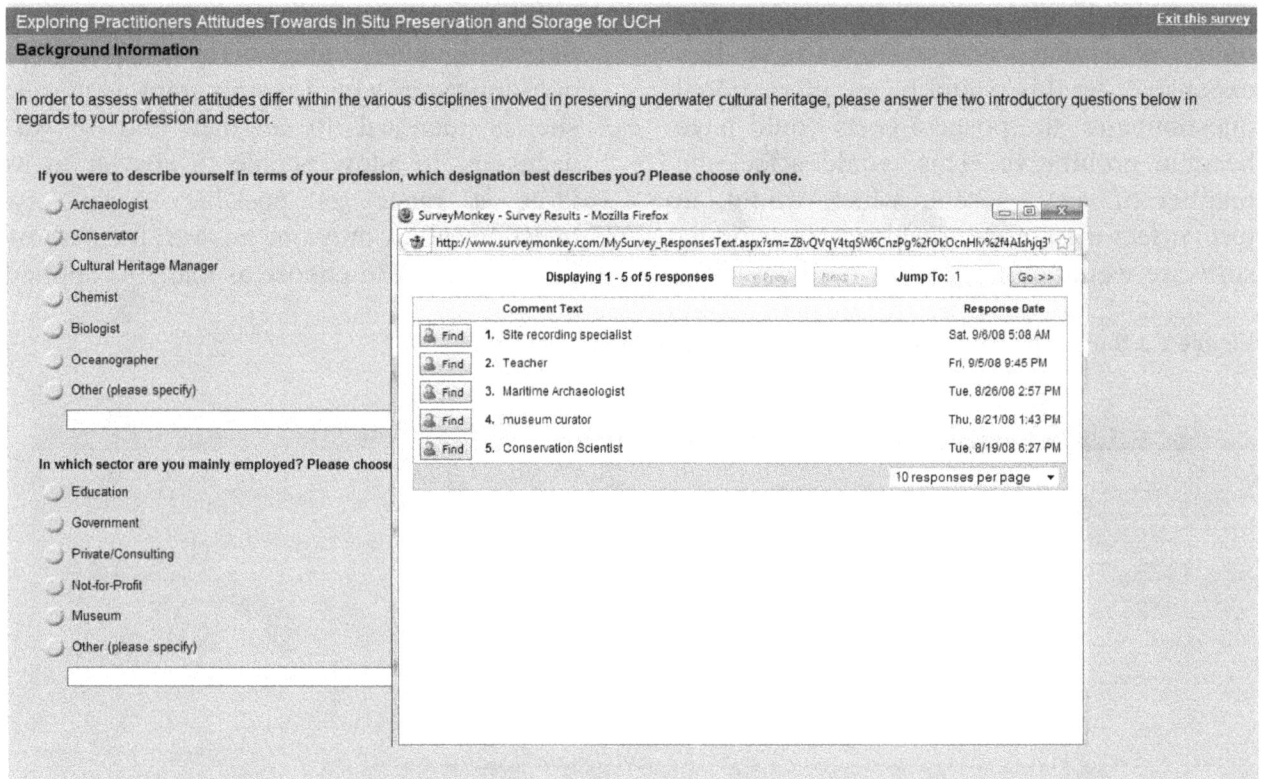

Figure 2-3 Box showing "Other" responses (SurveyMonkey™).

copies.

The analysis tool provided by SurveyMonkey™ is relatively simple. It provides basic information about the number of respondents who answered a question as well as the count and percentage for each response. Bar graphs present visual cues (Figure 2-2). All questions allowing the respondent to provide their own response are viewable in a separate window (Figure 2-3). As a more in-depth analysis is desirable, a separate statistical analysis programme, SPSS™ Statistics 17.0 (formerly Statistics

removed and stored elsewhere.

The appeal of SPSS™ for those who are not statisticians is that the user does not have to perform any complicated mathematics or understand complex formulas. The software completes the statistical calculations. The researcher has only to understand the data, the type of results each calculation is meant to produce and how to interpret them. Outputs include tables and a variety of graphs. All can be exported into a number of programmes suited to presentation and publishing.

As this questionnaire is inherently qualitative rather than quantitative, responses will be transformed into numbers as a way of ordering the data for processing by the software. This type of data is known as nominal data (Argyrous 2000:10). Nominal data limits the types of analysis performed to those that can be executed on single low-level data. This does not, however, preclude thorough analysis. What it does mean is that written responses by respondents need to be treated differently. In most instances they will receive two analyses: one as a group entitled "Other" which allows certain numerical concepts such as percentages to be applied and a second outside the statistics programme by a qualitative review similar to that used in the literature reviews. The primary form of analysis used on the questionnaire data will be descriptive statistics, including single variable frequencies and bar graphs; multiple response frequencies and bar graphs; and bivariate and multivariate measures of association such as Cramer's Vs, lambdas and chi squares (Argyrous 2000:38-39).

The ultimate aim of this research is to explore practitioners' attitudes towards *in situ* preservation and storage for underwater cultural heritage in light of the existing body of knowledge. The final step of the methodology aims to examine the implications of the questionnaire and literature review by combining and analysing both end products in order to underline current trends and patterns. *In situ* preservation and storage is an important device in the tool kit of maritime archaeologists and submerged cultural heritage managers. *In situ* preservation and storage needs to be understood in terms of its definitions and capabilities. It is believed that the proposed method of research has the ability to highlight trends in current thought and practice and provide a comprehensive overview of the types of scientific and archaeological enquiries being made into *in situ* preservation and storage. The most common forms of *in situ* preservation and storage will be identified along with the reasoning behind practitioners' choices for excavation, preservation, storage and monitoring. From this, recommendations for the direction of future studies should become apparent. Evaluation of the methods used to gather subject matter will hopefully highlight the validity of the results and advance new hypotheses for further exploration. While by no means definitive, this research should, through the combination of a literature review and a practitioners' questionnaire, provide a general base from which future studies can be conducted.

Acknowledgements

I would like to thank the following: the Flinders University Department of Archaeology and Mark Staniforth; my supervisors, Jennifer McKinnon and Vicki Richards, for their support and guidance; my colleague, Peter Ross, for assisting with html programming for the online questionnaire; Maarten van Oort, for assistance with appropriate language for English second-language participants; Alice Ortmann, for participating in "tense" discussions as well as for testing online delivery; and Laura Ortmann van Oort for her editing skills and help with the questionnaire from start to online delivery.

References

Alreck, P. L. and R. B. Settle

2004   *The Survey Research Handbook,* 3rd edition, McGraw-Hill/Irwin, Boston, MA.

Argyrous, G.

2000   *Statistics for Social and Health Research with a Guide to SPSS*, Sage, London, England.

Babits, L. E. and H.Van Tilburg (editors)

1998   *Maritime Archaeology: A Reader of Substantive and Theoretical Contributions*, Plenum Press, New York.

Bergstrand, T. and I. Nyström Godfrey (editors)

2007   *Reburial and Analyses of Archaeological Remains: Studies on the Effect of Reburial on Archaeological Materials Performed in Marstrand, Sweden 2002-2005, The RAAR Project*, Bohusläns Museum and Studio Västvensk Konservering, Uddevalla, Sweden.

Björdal, C. G., G. Daniel and T. Nilsson

2000   Depth of Burial: An Important Factor in Controlling Bacterial Decay of Waterlogged Archaeological Poles, *International Biodeterioration and Biodegradation* 45(1-2):15-26.

Björdal, C. G. and T.Nilsson

1999   Laboratory Reburial Experiments, In *Proceedings of the 7th ICOM-CC Working Group on Wet Organic Archaeological Materials Conference, Grenoble, France*, C. Bonnot-Diconne, X. Hiron, Q. Khoi Tran and P. Hoffman, editors, pp. 71-77, Arc-Nucléart, Grenoble, France.

Björdal, C. G. and T. Nilsson

2002   Waterlogged Archaeological Wood—A Substrate for White Rot Fungi During Drainage of Wetlands, *International Biodeterioration and Biodegradation* 50(1):17-23.

Björdal, C. G. and T. Nilsson

2008   Reburial of Shipwrecks in Marine Sediments: A Long-Term Study on Wood Degradation, *Journal of Archaeological Science* 35(4):862-872.

Björdal, C. G., T. Nilsson and G. Daniel

1999   Microbial Decay of Waterlogged Archaeological Wood Found in Sweden Applicable to Archaeology and Conservation, *International Biodeterioration and Biodegradation* 43(1-2):63-73.

Björdal, C. G., T. Nilsson and R. Petterson

2007   Preservation, Storage and Display of Waterlogged Wood and Wrecks in an Aquarium: "Project Aquarius", *Journal of Archaeological Science* 34(7):1169-1177.

Caple, C.

1998   Parameters for Monitoring Anoxic Environments, In *Preserving Archaeological Remains In Situ*, M. Corfield, P. Hinton, T. Nixon and M. Pollard, editors, pp. 113-123, Museum of London Archaeological Service and University of Bradford, London, England.

Caple, C., D. Dungworth, and P. Clogg

1997   Results of the Characterisation of the Anoxic Waterlogged Environments Which Preserve Archaeological Organic Materials, In *Proceedings of the 6th ICOM Group on Wet Organic Archaeological Materials Conference*, P. Hoffmann, T. Grant, J. A. Spriggs and T. Daley, editors, pp. 57-71, Ditzen, Bremerhaven, Germany.

Cederlund, C.O. (editor)

2004   *Monitoring, Safeguarding and Visualizing North-European Shipwreck Sites –
Challenges for Cultural Resource Management: Final Report*, The National Board of
Antiquities, Helsinki, Finland.

Corfield, M.

1996   Preventive Conservation for Archaeological Sites, In *Archaeological Conservation and
Its Consequences: Preprints of the Contributions to the Copenhagen Congress, 26-30
August 1996*, A. Roy and P. Smith, pp. 32-37, International Institute for Conservation of
Historic and Artistic Works, London, England.

Creswell, J. W.

2003   *Research Design: Qualitative, Quantitative, and Mixed Methods Approaches*, 2nd
edition, Sage Publications, Thousand Oaks, California.

Creswell, J. W. and V. L. Plano Clark

2007   *Designing and Conducting Mixed Methods Research*, Sage, Thousand Oaks, California.

De Vaus, D. A.

2002   *Surveys in Social Research*, 5th edition, Allen and Unwin, Routledge, St. Leonards,
New South Wales.

Delgado, J. (editor)

1997   *Encyclopaedia of Underwater and Maritime Archaeology*, British Museum Press,
London, England.

Desor, E.

1865   *Les Palafittes ou Constructions Lacustres du Lac de Neuchatel*, n.p., Paris, France.

Fink, Arlene.

2005   *Conducting Research Literature Reviews: From the Internet to Paper*, 2nd edition, Sage
Publications, Thousand Oaks, California.

Foddy, W. H.

1988   *Elementary Applied Statistics for the Social Sciences*, Harper and Row, Sydney, New
South Wales.

Foddy, W. H.

1993   *Constructing Questions for Interviews and Questionnaires: Theory and Practice in
Social Research*, Cambridge University Press, Cambridge, England.

Godfrey, I., E. Reed, V. Richards, N. West and T. Winton

2005   The *James Matthews* Shipwreck - Conservation Survey and *In-Situ* Stabilisation, In
*Proceedings of the 9th ICOM Group on Wet Organic Archaeological Materials
Conference*, P. Hoffmann, K. Strætkvern, J. A. Spriggs and D. Gregory, editors, pp. 40-
76, ICOM Committee for Conservation Working Group on Wet Organic
Archaeological Materials, Bremerhaven, Germany.

Goodburn-Brown, D. and R. Hughes

1996   A Review of Some Conservation Procedures for the Reburial of Archaeological Sites in
London, In *Archaeological Conservation and Its Consequences: Preprints of the
Contributions to the Copenhagen Congress, 26-30 August 1996*, A. Roy and P. Smith,
pp. 65-69, International Institute for Conservation of Historic and Artistic Works,
London, England.

Green, J. N.

2003 *Maritime Archaeology: A Technical Handbook,* 2nd edition, Elsevier Academic, San Diego, California.

Hart, C.

1998 *Doing a Literature Review: Releasing the Social Science Research Imagination,* Sage, London, England.

Heldtberg, M., I. MacLeod and V. Richards

2005 Corrosion and Cathodic Protection of Iron in Seawater: A Case Study of the *James Matthews* (1841), In *Metal 04: Proceedings of the International Conference on Metals Conservation,* J. Ashton and D. Hallam, editors, pp. 75-87, National Museum of Australia, Canberra, ACT.

Holden, J., L. J. West, A. J. Howard, E. Maxfield, I. Panter and J. Oxley

2006 Hydrological Controls of *In Situ* Preservation of Waterlogged Archaeological Deposits, *Earth-Science Reviews* 78(1-2):59-83.

Hosty, K.

1988 Bagging the *William Salthouse*: Site Stabilization Work on the *William Salthouse, Bulletin of the Australian Institute for Maritime Archaeology* 12(2):13-16.

International Council on Monuments and Sites

1996 Charter on the Protection and Management of Underwater Cultural Heritage (1996), International Council on Monuments and Sites, ICOMOS International Secretariat, Paris, France <http://www.international.icomos.org/charters/underwater_e.pdf>. Accessed 26 May 2008.

MacLeod, I.

1995 *In Situ* Corrosion Studies on the Duart Point Wreck, 1994, *International Journal of Nautical Archaeology* 24(1):53-59.

MacLeod, I.

1993 Metal Corrosion on Shipwrecks: Australian Case Studies *Trends in Corrosion Research* 1:221-245.

MacLeod, I.

1996 *In Situ* Conservation of Cannon and Anchors on Shipwrecks Sites, In *Archaeological Conservation and Its Consequences: Preprints of the Contributions to the Copenhagen Congress, 26-30 August 1996,* A. Roy and P. Smith, pp. 111-115, International Institute for Conservation of Historic and Artistic Works, London, England.

MacLeod, I.

1998 *In-Situ* Corrosion Studies on Iron and Composite Wrecks in South Australian Waters: Implications for Site Managers and Cultural Tourism, *Bulletin of the Australian Institute for Maritime Archaeology* 22:81-90.

MacLeod, I., P. Morrison, V. Richards and N. West

2005 Corrosion Monitoring and the Environmental Impact of Decom Naval Vessels as Artificial Reefs, In *Metal 04: Proceedings of the International Conference on Metals Conservation,* J. Ashton and D. Hallam, editors, pp. 53-87, National Museum of Australia, Canberra, ACT.

Manders, M. R.

2004   Protecting Common Maritime Heritage. The Netherlands Involved in Two EU-Projects: MoSS and BACPOLES, In *Tutela, Conservazione e Valorizzazione del Patrimonio Culturale Subacqueo,* pp. 279-291, Massa Editore, Napoli, Italy.

McCann, A. M.

2001   An Early Imperial Shipwreck in the Deep Sea off Skerki Bank, *Rei Cretariae Romanae Fautorum Acta 37,* pp. 257-264.

McCann, A. M. and J. P. Oleson

2004   Deep-Water Shipwrecks off Skerki Bank: The 1997 Survey, *Journal of Roman Archaeology,* Supplementary Series No. 58.

McCarthy, M.

1987   The SS *Xantho* Project: Management and Conservation In *Conservation of Wet Wood and Metal,* I. MacLeod, editor, pp. 9-15, Western Australian Museum, Perth, Western Australia.

Moran, V.

1997   Some Management Options for the Perched Hull *Day Dawn, Bulletin of the Australian Institute for Maritime Archaeology* 21(1-2):129-132.

Muckelroy, K.

1998   Introducing Maritime Archaeology. In *Maritime Archaeology: A Reader of Substantive and Theoretical Contributions,* L.E. Babits and H. Van Tilburg, editors, pp. 23-38, Plenum Press, New York.

Nash, M.

1991   Recent Work on the Sydney Cover Historic Shipwreck, *Bulletin of the Australian Institute for Maritime Archaeology* 15(1):37-47.

The Nautical Archaeology Society

1992   *Archaeology Underwater: The NAS Guide to Principles and Practice,* 1st edition, Nautical Archaeology Society and Archetype Publications, London, England.

Oxley, I.

1998   The *In-Situ* Preservation of Underwater Sites, In *Preserving Archaeological Remains in Situ,* M. Corfield, P. Hinton, T. Nixon and M. Pollard, editors, pp. 159-173, Museum of London Archaeological Service and University of Bradford, London, England.

Stewart, J., L. D. Murdock and P. Waddell

1995   Reburial of the Red Bay Wreck as a Form of Preservation and Protection of the Historic Resource, In *Material Issues in Art and Archaeology: IV,* P. V. Vandiver, J. R. Druzik, J. L. G. Madrid, I. C. Freestone and G. S. Wheeler, editors, pp. 791-806, Materials Research Society, Pittsburgh, Pennsylvania.

Ucelli, G.

1950   *Le Navi di Nemi.* La Libreria dello Stato, Rome, Italy.

United Nations Educational, Scientific and Cultural Organization

2001   *UNESCO Convention on the Protection of the Underwater Cultural Heritage.* UNESCO World Heritage Centre, United Nations, Paris, France <http://unesdoc.unesco.org/images/0012/001246/124687e.pdf#page=5>. Accessed 28 May 2008.

Winton, T. and V. Richards

2005   *In-Situ* Containment of Sediment for Shipwreck Reburial Projects, In *Proceedings of the 9th ICOM Group on Wet Organic Archaeological Materials Conference*, P. Hoffmann, K. Strætkvern, J. A. Spriggs and D. Gregory, editors, pp. 77-89, ICOM Committee for Conservation Working Group on Wet Organic Archaeological Materials, Bremerhaven, Germany.

3   Conserving the WWII Wrecks of Truk Lagoon

Jon Carpenter
Department of Materials Conservation, Western Australian Museum, Shipwreck Galleries, 45-47 Cliff Street, Fremantle, Western Australia, 6160

Ian MacLeod
Collection Management and Conservation, Western Australian Museum, Locked bag 49, Welshpool DC, Western Australia, 6986

Vicki Richards
Department of Materials Conservation, Western Australian Museum, Shipwreck Galleries, 45-47 Cliff Street, Fremantle, Western Australia, 6160

As an ally of the West in WWI Japan received a mandate from the League of Nations to occupy and govern German colonial possessions in Micronesia. Subsequently, the atoll known as Chuuk Lagoon (Truk) was secretly developed to become Japan's principal naval base in the Pacific. Japanese aggression during WWII resulted in Truk lagoon becoming a target for US naval air forces and consequently a number of Japanese ships were sunk and defending aircraft shot down. These ships and aircraft are considered significant historical remains from WWII. The wreck sites are the main tourist attraction and a primary industry for present day Truk. Investigation of these sites aims to document the archaeological evidence and ascertain the structural integrity of the wrecks for management purposes. The principal investigator for this project is Maritime Archaeologist Dr Bill Jeffery, supported by a team of conservation specialists, marine biologists and EarthWatch volunteers. This paper primarily discusses the conservation strategy and methodology applied during investigations of the ship and aircraft wreck sites with some preliminary results.

Introduction

As an ally of the West in WWI Japan received a mandate from the League of Nations to occupy and govern German colonial possessions in Micronesia. Subsequently the atoll known as Truk Lagoon (Chuuk) was secretly developed to become Japan's principal naval base in the Pacific. Japanese aggression during WWII resulted in Truk lagoon becoming a target for US naval air forces and consequently a number of their ships were sunk and defending aircraft shot down. These ships and aircraft are considered significant historical remains from WWII. The wreck sites are the main tourist attraction and a primary industry for present day Truk. Investigation of these sites aims to document the archaeological evidence and ascertain the structural integrity of the wrecks for management purposes. This paper primarily discusses the conservation strategy and methodology applied during investigation of the ship and aircraft wreck sites with some preliminary results.

Historical background

Chuuk, more popularly known as Truk Lagoon, is located in the Federated States of Micronesia (Figure 3-1). Micronesia became a German colonial possession in 1899. Allied to Britain, France and the US during WWI, Japan declared war on Germany in August 1914. In October of the same year Japan captured Truk supported by the cruiser HIJMS *Kurama*. Despite concerns by the US, Japan received a mandate from the League of Nations to govern occupied Micronesia provided the islands were de-militarized. Regardless of this restriction Japan began developing a strong military presence and in great secrecy Truk was established as their principal naval base in the Pacific. The base was strongly fortified being protected by many anti-aircraft guns and large coastal guns some of which were mounted inside caves cut by forced labour.

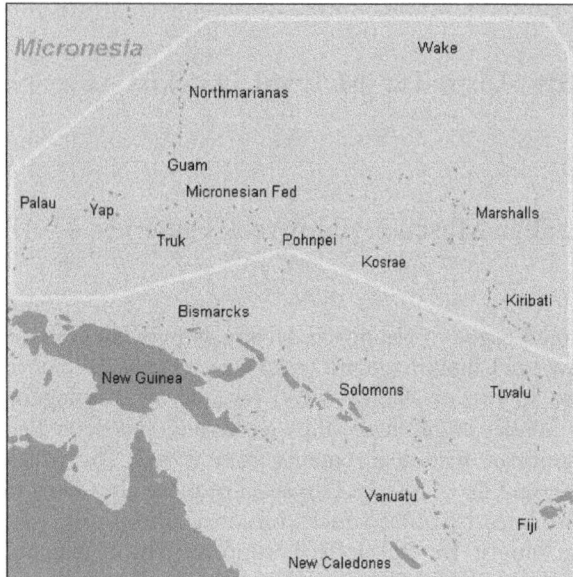

Figure 3-1 Micronesia (Federated States of Micronesia.org).

As a result of Japanese aggression during WWII Truk atoll became a target for US naval air forces. Consequently many of Japan's ships were sunk and a number of defending aircraft were shot down over the lagoon. US Operation Hailstone, in February 1944, was particularly significant as this deployment resulted in considerable losses to the Japanese with some 45 ships sunk and more than 270 aircraft destroyed.

The Truk Lagoon Project

Maritime Archaeologist Dr Bill Jeffery, with sanction and support from the Historic Preservation Office (HPO) in Truk, has been undertaking an investigation of the submerged remains of the ships and aircraft in the lagoon. The project team includes conservation specialists, marine biologists and EarthWatch volunteers. Conservation is an important aspect of this study and is the primary concern of personnel from the

Department of Materials Conservation, Western Australian Museum. Interpretation of data acquired will help to determine the structural condition of the wrecks and provide an indication of their structural life expectancy. Besides the scientific knowledge gained this will also provide information that is important for the implications it will have on the cultural and tourism resource.

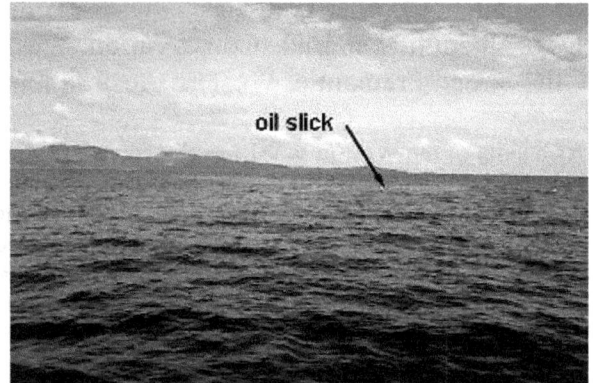

Figure 3-2 Oil slick on the sea surface (Author).

A primary aim of the project is to preserve the sites through managerial intervention. Besides the obvious risk to SCUBA diving visitors, should any part of a wreck collapse, there is the potential for a wider negative impact if large quantities of hydrocarbons, principally fuel oil trapped in the ships, is released (Figure 3-2). Interpretation of the corrosion survey data will provide a good indication of the condition and therefore structural integrity of the wrecks. The information will permit recommendations to be made on measures to control their deterioration. On a smaller scale, conservation advice is available for wreck artefacts (i.e. those objects recovered prior to laws forbidding their collection or for more recently confiscated artefacts).

The overall aims and objectives of the project are to:
- Document the archaeological evidence
- Monitor the structural integrity of the wrecks

34

- Record the biological diversity
- Provide practical guidelines for the Historic Preservation Office (HPO) to conduct regular surveys and monitor the status of the sites to facilitate preservation
- Provide conservation advice and training to preserve recovered artefacts

Conservation Strategy

The conservation strategy has been outlined to:
- Determine the mechanisms and rates of corrosion
- Ascertain the status of the wrecks in terms of structural integrity
- Attempt to predict structural lifespan
- Make recommendations to implement measures to attempt to stabilize the sites

The following is a table of the wrecks and aircraft that have so far been investigated for the corrosion study. (Refer to Figure 3-3 for positions of the shipwrecks and aircraft).

Table 3-1 Ships with corrosion data (Author).

Ships and Aircraft
14. *Hoyo Maru*
16. *Susuki*
18. *Nippo Maru*
30. *Fujikawa Maru*
32. *Gosei Maru*
34. *Yubae Maru*
38. *Hino Maru*
40. *Sankisan Maru*
40. *Shinkoku Maru*
40. *Sapporo Maru*
40. *Eisen*
B. Betty (Mitsubishi G4M)
E. Emily (Kawanishi H8K)
Z. Zeke or Zero (Mitsubishi A6M Reisen)
JY. Judy (Yokosuka D4Y Suisei)

The Corrosion Process

Metal submerged in seawater corrodes in a micro-environment that prevents the formation of protective oxide layers normally associated with terrestrial corrosion processes. The principle factors that control corrosion include: water movement, dissolved oxygen, salinity, temperature and metal composition. Marine growth, accretions and concretions separate metal surfaces from direct access to dissolved oxygen. These overlying materials also act as semi-permeable membranes allowing the concentration of corrosive chloride salts to increase and acid conditions to develop beneath the layers. After a few

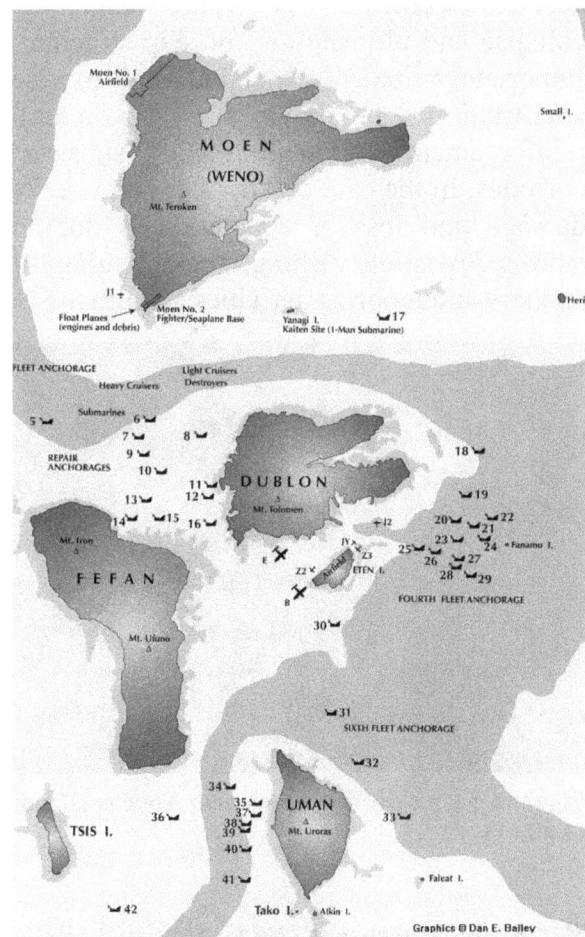

Figure 3-3 Location of ships and aircraft (Bailey 2001).

35

years of submergence in the marine environment the corrosion rate beneath the concretion will fall to a steady state value. For example, during procedures to record corrosion data from the bridge structure of an unidentified wreck (Figure 3-4) it was found to be extensively weakened due to corrosion. However concretion and marine encrustation is reinforcing this unstable structure.

Iron corrosion products stimulate marine growth; consequently iron ships are characterized by a diverse, thickly encrusting and variable layer of marine biota. If this layer is damaged it can result in accelerated corrosion, premature weakening, collapse and ultimate loss of ship structure. Importantly intact concretion and marine encrustation provides structural reinforcement as the metal substrate corrodes. In the case of Truk Lagoon wrecks damage and loss of concretion is due to storms, dynamite fishing, the dragging of anchors and mooring activities (Figure 3-5).

Figure 3-5 Unidentified wreck near Tonoas Island (Author).

Recording Corrosion Data

To record corrosion data underwater a pH meter and multimeter, housed in a water-tight housing are used. External probes connected to these instruments permit a series of pH and corrosion potential measurements to be made. Accordingly several WWII shipwrecks and the wrecks of aircraft have been probed to determine the underlying nature of the corrosion processes. Dissolved oxygen concentration, salinity and temperature profiles were also recorded from the water column over each site to enable predictions of corrosion rates to be made with a reasonable degree of certainty.

Figure 3-4 Mooring damage (concretion stripping) *Gosei Maru* (Author).

The methodology for recording the pH and corrosion potential requires that a pneumatic drill, fitted with a masonry type bit be used. An access hole is drilled into the layer of concretion and corrosion products until contact with solid metal is made. In a closely coordinated manoeuvre a pH probe is inserted into the hole immediately after the drill bit is removed to minimize the ingress of seawater (the diameter of the drill bit, 16 mm, is only slightly more than that of the pH probe). Once it has stabilized, the pH reading is recorded and the probe is removed. A platinum electrode is then inserted to make contact with the surface of the metal core and record the corrosion potential. Corrosion activity generates a small measurable current, interpretation of this, in conjunction with the pH reading determines how actively or not the metal is corroding. The depth of the hole is recorded and then to prevent an increase in corrosion activity at the drill site it is plugged and sealed with waterproof epoxy putty (Selleys Aqua Knead It). Photographic images recorded at each drill site enable each location to be revisited for future data acquisition.

As water depth increases, the pH of surfaces becomes less acid; correspondingly the corrosion rate of metal is expected to diminish with increasing water depth. The corrosion rate minimum is reached around a depth of 27 m. Environmental factors / influences aside, the corrosion rates in wrecks are not always consistent due to the variable combinations of materials present, particularly the relationship between the different metals and the individual components/impurities that make up a metal. The average corrosion rate for iron in seawater has been determined to be 0.1 mm per year.

The varying site locations, depth and orientation of the wrecks provide a wide range of conditions in which to assess corrosion rates. The corrosion rates of the wrecks in Truk Lagoon are lower than iron shipwrecks at the same depth in open ocean waters, which demonstrates the protective nature of the sheltered waters of the lagoon (MacLeod 2002). As an example, the high profile position of the ship guns mounted on the *Fujikawa Maru* has resulted in them being more corroded due to greater exposure to water movement and a higher level of dissolved oxygen. The more exposed starboard side of the *Gosei Maru* is corroding more actively than the deeper partly buried port side (MacLeod 2002). The metal plates around the torpedo breeched hulls of ships like the *Hoyo Maru* are anticipated to exhibit more extensive corrosion due to the stresses created in the metal. On a more localized scale the interaction between the different metals associated with ships fixtures and cargo materials will also influence corrosion

activity; it is usual for the more noble metals to be protected when iron is in contact with them.

Figure 3-6 Sponges and other marine growth on the undersurfaces of the *Emily* Floatplane wing (Author).

Aluminium Corrosion

As in air the exposed aluminium surfaces of an aircraft in a marine environment are usually preserved by a passive oxide film that protects the alloy from rapid corrosion. Aluminium corrosion products are considered biologically inert; in theory this should mean that surfaces would be easily colonized by marine biota. In actuality the opposite is generally the case with large areas of aluminium surfaces remaining free from colonization. Exceptions occur in areas where aluminium surfaces are more protected and subject to lower light levels such as in under-wing surfaces and within the interior spaces of aircraft (Figure 3-6). This implies that exposure of aluminium to more dynamic water movement, likely to include particulates that would impinge on it, effectively scours the surface and prevents marine biota from achieving a firm and lasting attachment on what is a very smooth surface.

Localized marine growth does occur on aluminium surfaces exposed to more dynamic conditions but this may be due to the presence of materials other than aluminium, such as iron and/or areas subjected to damage. In this case partial encrustation of aluminium will tend to set up differential aeration cells and so cause nearby metal to suffer from accelerated pitting and general corrosion. Any light pitting corrosion that is evident on the sheet aluminium surfaces in the marine environment can be a consequence of impurities and/or due to the deliberate

Figure 3-7 Recording pH and corrosion data on the *Emily* wing (Author).

inclusion of other metals such as copper.

The variable composition and the likelihood of the decreasing quality of the aluminium alloys used in Japanese WWII aircraft, as the supply of war materials diminished, makes a study of the corrosion mechanisms more complex. Higher magnesium and iron content leads to more corrosion. Some equipment in the aircraft can be cathodically protected, such as the steel guns, and in this situation the more reactive aluminium will corrode. The overall variety, composition and relationship between the different materials used in the manufacture of aircraft will determine corrosion processes. Zero (Zeke) aircraft have a central spar made of a zinc/chrome alloy for example.

Recording pH of Aluminium

The recording of pH on some aluminium surfaces is difficult owing to the absence or presence of only a very thin layer; less than 1 mm of marine growth and corrosion products (Figure 3-7). Consequently pH values on aircraft are generally very conservative (i.e. the underlying acidity is anticipated to be higher than reported).

Additional Data

Additional information was gathered from these sites which aids in the assessment of corrosion measurements. These include:

Weather, Sea Conditions, Swell, currents and Tidal Information. These influences affect oxygen availability and thereby directly impact corrosion rates. Disturbance of sand and sediments inhibits marine growth, erodes protective concretions or prevents their formation. Exposed metals are sandblasted and corrosion is accelerated.

Water Temperature. The influence of temperature on biological growth and consequent encrustation affects corrosion rates. Generally a rise of 10°C in water temperature doubles the corrosion rate.

Water Depth to Wreck (min/max). Generally deeper sites have reduced water movement and oxygen availability. Colder temperatures slow the corrosion rate.

Underwater Visibility. In poor visibility reduced light penetration affects the establishment of marine growth and consequently corrosion rates. If SCUBA diver visitation is reduced because of poor underwater visibility, then minimization of direct human disturbance on the site may be beneficial.

Distance and Orientation from Land/Reef. Protection afforded to a wreck site by nearby land or reef has effect on water movement and therefore corrosion rates. Close proximity to land also has potential for human disturbance, pollution and freshwater effects on corrosion rates.

Evidence of active corrosion. Typical prominent orange, red/brown (rust) patches or spots indicate active iron corrosion. Corroding areas on aluminium can be more difficult to identify until it has become perforated. Fluffy white/grey patches and pustules may develop. If copper is present, as in Duralum, then typical blue/green copper corrosion products may also be evident.

Dominant Encrusting Organisms on Surface. This need only be a very general survey, photographically documenting the main encrusting organisms and recording any evidence of dynamite fishing and/or storm damage.

Dynamite and Storm Damage. Recent damage caused to submerged sites by dynamite fishing and after severe storms is easily identified by the large areas of exposed metal, detached concretion and areas devoid of secondary marine growth. Often the metal will show signs of active corrosion. It is important that these damaged areas are accurately documented in an initial survey. Information gathered on any subsequent surveys can be directly compared to this baseline thus revealing any changes in the corrosion activity, extent of colonisation, etc.

Conclusion

It is predicted that several wrecks in Truk Lagoon will undergo significant collapse in the next 10-15 years (MacLeod 2002). To slow the deterioration of the shipwrecks in Truk Lagoon it is essential that all dynamite fishing be stopped and mooring activities must be better managed to prevent damage to the protective concretion and marine life. Regular corrosion data acquisition needs to continue in order to obtain more and consistent corrosion information to provide a clearer picture of the processes involved. This will improve the accuracy of predictions concerning the structural integrity of the wrecks. Funding must also be sought to ensure it will continue. Corrosion control measures need to be investigated and cathodic protection methods should be trialled. The release of hydrocarbons such as fuel oil is an important concern. If corrosion control measures can be implemented it may delay more extensive perforation of fuel storage tanks in the wrecks, but pro-active measures still should be put in place to remove the oil. At the very least oil recovery equipment should be on-standby with an action plan already prepared to deal with and contain any major release of oil. The thicknesses of the residual metal in hull plates and oil storage tanks (if possible) need to be measured to help determine current status and predicted life expectancy.

Acknowledgements

Our thanks to Dr Bill Jeffery for the opportunity to contribute to the Truk project. Thanks to Dan E. Bailey for permission to use illustrations from his book *World War II Wrecks of the Truk Lagoon.*

References

Bailey, D. E.
2001   *World War II Wrecks of the Truk Lagoon*, North Valley Diver Publications, Redding, California.
MacLeod, I.
2002   *Metal Corrosion in Chuuk Lagoon; A Survey of Iron Shipwrecks and Aluminium Aircraft*, Western Australian Museum, Freemantle, South Australia.

4 Cegrass™, Sand and Marine Habitats: A Sustainable Future for the William Salthouse Wreck

Hanna Steyne
Heritage Victoria, GPO Box 2392, Melbourne, Victoria, 3001

William Salthouse is one of Victoria's oldest and most intact shipwrecks. Shortly after its discovery, the site was looted and suffered rapid erosion. Initial attempts at *in situ* stabilisation failed, but the placement of artificial Cegrass™ on the site was an immediate success. The site has been stable for the past 12 years; however a dive in September 2008 unexpectedly discovered deep scours in the sandbank. The current condition of the site has been assessed in relation to changing local environments and the continued use of a permit only access system for divers to the site.

Introduction

William Salthouse was built in Liverpool in 1824 as a two-masted brig of 251 tons, and was used initially as a trader between Britain and the West Indies, and later to the East Indies. After a change in ownership, *William Salthouse* was sent on a voyage from Britain to the British Dominion of Canada and on to the British Colonies in Australia in 1841. This voyage was the first time a British trading vessel had come directly from British North America to Australia, thereby flouting British Navigation Laws which prevented direct trading between the British Colonies at the

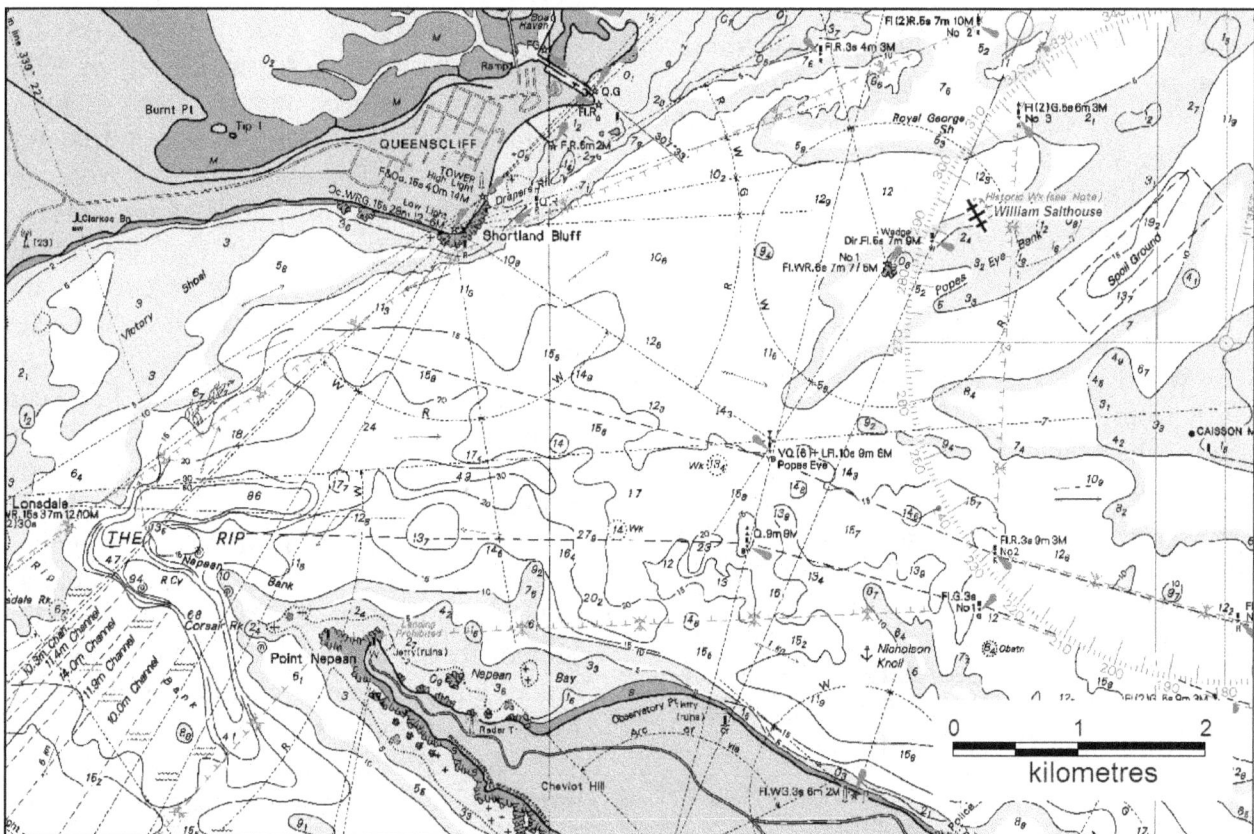

Figure 4-1 Location of the *William Salthouse* wreck in southern Port Phillip Bay, Victoria (AUS00158 - Entrance to Port Phillip 2002).

41

time. *William Salthouse* attempted to enter Port Phillip Bay on 27 November 1841 with a mixed cargo destined for Melbourne, but struck a rock off Point Nepean, which unshipped the rudder. Although the pilot attempted to sail the vessel onwards, it continued to take on water and was run ashore on the sand bank known as Pope's Eye. By morning *William Salthouse* had sunk, with two meters of water over the deck (Staniforth and Vickery 1984) (Figure 4-1).

Discovery

The site was discovered in August 1982 by sports divers and the location of the wreck soon became known within the diving community. The shipwreck was sitting almost upright on the seabed, with a slight list to starboard and intact to the deck level (Staniforth and Vickery 1984). The site quickly became the focus of activities by souvenir hunters who inflicted considerable damage to the site as they unsuccessfully hunted for 'treasure.' The site was reported to the Victoria Archaeological Survey (VAS), the pre-cursor to the historical and maritime heritage units at Heritage Victoria, in December 1982 after the finders saw the damage being done to the site by unscrupulous divers (Harvey 1996).

When the site was first examined by staff from VAS, shipwreck material was seen to be scattered up to 50 m away from the main wreck a result of both the wrecking process and disturbance by divers. Despite this, a considerable part of the hull and its contents were still intact. The main site measured approximately 25 m in length and 8 m wide, and barrels of cargo were visible *in situ* with straw packing in place (Staniforth and Vickery 1984).

Looting and Protection

The site was designated an Historic Shipwreck under the Victorian *Historic Shipwrecks Act* 1981 on 22 December 1982. This allowed divers to access the site but prohibited interference, damage and the removal of artefacts. Despite the declaration, further damage was reported in January 1983, which led to the declaration of a 250 m radius Protected Zone around the site on 9 February 1983. This declaration prohibited access to the site without a permit and put a stop to looting activities (Strachan 1988).

Erosion and Control

Whilst damage caused directly by looting activities was stopped by the implementation of the Protected Zone, it became clear that the site had become destabilised, and the sand which had preserved and covered the site up until 1982 was continuing to erode away. As the erosion continued, VAS mobilised in 1983 undertaking surveys, the recovery of loose artefacts and excavating two test trenches across the site. The results of this work are described in Staniforth and Vickery 1984.

The survey identified that the strong tidal currents in the area were causing the scouring and erosion (Hosty 1989); however the initial trigger leading to the site's exposure from the sandbank has not been established. No obvious changes in the local environment, such as storms or dredging, are recorded to have taken place around this time. It is possible that unreported channel works or damage to the sandbank by scallop dredgers took place, as it is unlikely that the previously stable sandbank would suddenly erode without some change to the local hydrodynamic environment.

The excavation and survey by VAS attracted extensive press coverage and diver interest, which brought increasing calls from the public to open the site to visiting divers.

As the site seemed to stabilise a little, the site was re-opened to divers on a permit system in 1984. The permits limited the number of divers allowed to visit the site at one time to 12, but the scheme was hugely popular and hundreds of divers visited the site each year. With such large numbers of divers visiting the site through the permit system, the rates of erosion on the site began to increase again. Whilst most divers seemed to respect the site as a protected archaeological site, damage from divers, through poor buoyancy, hand fanning and moving of objects was seen (Harvey 1996).

Whilst permit access to the site continued, erosion of the site persisted despite attempts by VAS to institute control measures. The first of these was the placement early in 1985 of five small fences (Figure 4-2). The fences measured 1 m by 0.5 m and were intended to trap seaweed and sand, thereby building up sand around the site. However, by September 1985 fences were found to have had little effect (Harvey 1996).

Figure 4-2 One of the mesh fences placed on site in 1985 (Heritage Victoria 1985).

The second approach taken in an attempt to control and reverse the erosion, involved the placement of sediment on site. At first a hand dredge was used to fill the scour holes around the site, and when this failed a more approach was taken, dumping a full cargo of dredged spoil on to the site. This drastic action, however, had limited effect leaving only a light dusting of sand on the site (Harvey 1996).

Figure 4-3 Sandbags on the *William Salthouse* wreck supporting the hull structure (Heritage Victoria 1988).

With attempts at erosion control failing, the site was closed again to divers in June 1988. Sandbags were identified as the best short term measure to support the exposed hull sections, prevent further collapse, and buy time to investigate a long term solution (Hosty 1989). Six areas of exposed hull were sandbagged in 1988 (Figure 4-3). Details of this work can be found in Hosty 1989. The sandbags succeeded in temporarily preventing further collapse of the hull structure and reducing scouring. They also accumulated sand and marine growth around the wreck. By late 1989 however, the hessian bags had begun to degrade and sand was eroding away (Harvey 1996).

The Cegrass™ Plan

Artificial seagrass matting was identified as the only viable solution to site stabilisation, given both requirements from the Port of Melbourne Authority to use biodegradable materials for underwater construction, and the ethical frameworks laid out in the Burra Charter and ICOMOS Guidelines (1981) for non-intrusive, reversible and *in situ* conservation works. The Cegrass™ itself works by catching sediment amongst the grass like 'fronds', thereby facilitating the rebuilding of a sandbank. Cegrass™ had been developed for the Oil and Gas Industry in the North Sea, to prevent and control erosion around sub-sea structures. This was the first time it had been used on an archaeological site.

The 'grass' fronds themselves are made of polypropylene strips (1.6 cm wide), are buoyant and degrade when exposed to direct sunlight (which could happen if they broke and floated away from the site). The individual fronds are grouped in a clip, similar to natural seagrass, then the grouped grass fronds are attached to steel mesh sheets with 20 cm grid spacing. As the fronds themselves are buoyant, each sheet was weighted to ensure it stayed on the seabed (Figure 4-4). Details of the Cegrass™ deployment can be found in Harvey 1996.

Figure 4-4 Cegrass™ sheet in position on the *William Salthouse* wreck (Heritage Victoria 1990).

Three lengths of Cegrass™ were placed on the site in 1990 measuring 90 cm, 120 cm and 150 cm, with the longest fronds placed in areas of deepest scour. The arrangement of the Cegrass™ sheets on the seabed aimed to fill the scour holes and re-build a supporting sand dune. The top of the wreck was not covered with Cegrass™, but left exposed to enable visiting divers to see the remains of the site (Figure 4-5).

Cegrass™ Success

Regular site monitoring showed the Cegrass™ had almost immediate results, and over the next couple of years the sand build up and any new areas of erosion were carefully mapped. After just one week on the seabed there was 10-15 cm of sediment build up in the Cegrass™ fronds, whilst after six months on the seabed the scour holes had been filled and erosion of the wreck halted. Two new small scour holes were identified and filled by use of small Cegrass™ sheets, and it was noted that some sediment had also accumulated over the central part of the wreck in the two to three months after Cegrass™ was placed on site. Additional sheets of a light steel mesh were placed over the central areas without Cegrass™, to prevent loss of any loose artefacts and to provide a base for marine life to grow on. Over time, the steel mesh rusted away, and the last sheet was recorded in place in 1991.

Regular monitoring of the sand levels continued until 1993, when the site was deemed stable and re-opened to divers on a permit system. When Harvey published a review of the stabilisation works in 1996, the site was recorded as stable, with no problems relating to erosion. At this time the top 15-20 cm of Cegrass™ fronds were still visible above the sand and were heavily colonised by marine organisms (Harvey 1996).

Into the Twenty-first Century

Monitoring of the wreck of *William Salthouse* by Heritage Victoria has continued on almost an annual basis since 1996, and for ten years up until 2006, no noticeable changes in erosion levels have been seen. Figure 4-6 illustrates that in 2006 the Cegrass™ was mostly buried with the exposed tips covered with marine growth. Cargo barrels were exposed across the site, and many had bones (held in place by sand) visible inside them. Between 1996 and 2006 a small number of permits had been issued to divers. Interest from the diving community through this period varied depending on the level of media coverage of shipwrecks and the use of the site as a case study in AIMA/NAS training courses. Despite this variation, visitor numbers have remained reasonably small, in comparison to the early 1980s.

Today the wreck of *William Salthouse* is still within a Protected Zone (designated under the Victorian *Heritage Act* 1995), and access is still granted to divers under a permit system. No other access is allowed; however inspection in 2008 did reveal a concentration of beer cans and anchor damage on the starboard side of the wreck, suggesting illegal visits by fishers. The

numbers of divers visiting the site through the permit system is very small but steady and the permit system is well known to Victorian divers. In the past 12 months just four permits have been issued, one of which was a re-issue to a club who had bad weather for their scheduled dive. Each permit still allows only 12 divers on the site at a time. There does not seem to be any evidence of illegal diving, or of damage by divers. This suggests that education programs, such as the AIMA/NAS training courses, have been a success and that there is an increased appreciation of shipwrecks as fragile sites, rather than sources of trinkets.

Current Site Conditions

The wreck of *William Salthouse* lies with the bow at roughly 120 degrees

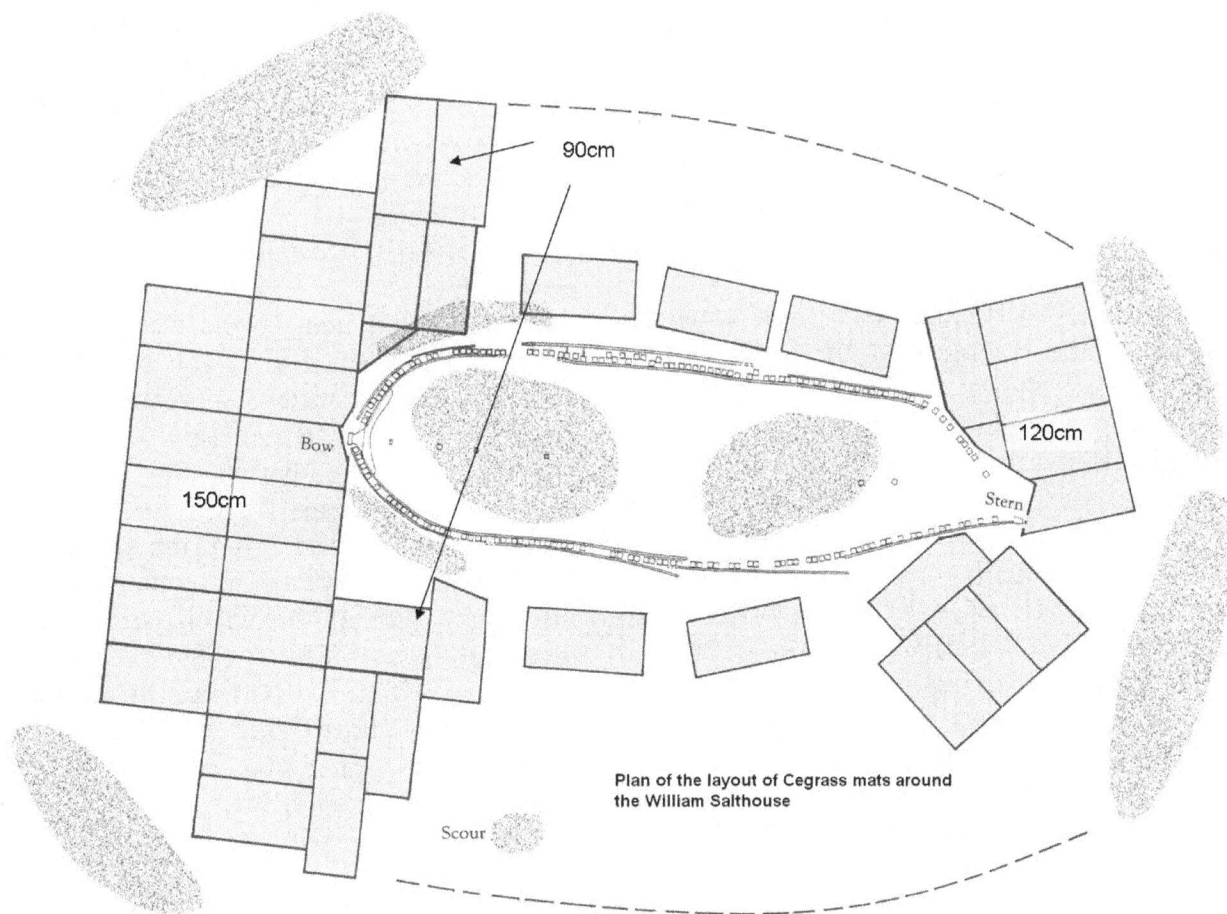

Figure 4-5 Position of the Cegrass™ mats around the *William Salthouse* wreck (Heritage Victoria 1990).

(Staniforth and Vickery 1984). The prevailing current runs across the site, with the stern more exposed to the flood tide, and the bow to the ebb. Tidal flows can reach 2.5 knots in the southern part of the bay and almost 7 knots over the Nepean Bank (Port of Melbourne Corporation 2004:13-3).

Figure 4-6 Cegrass™ buried, with heavy marine growth in 2006 (Hosty 2006).

The site was visited in September 2008 to undertake monitoring of the Cegrass™ and in order to report on the current condition of the site at the AIMA/ASHA conference in Adelaide. During this visit the exposed frond lengths around the site were measured and photographed and the results are presented in Figure 4-7. Significant changes in the condition of the sandbank were observed during the 2008 inspection, with areas of erosion and sand deposition quite different to those observed in 2006 and in previous years. Of particular concern was an area of deep scour in the supporting sandbank at the stern, where areas of pebble armour had developed, and the full length of the Cegrass™ (120 cm) was exposed down to the metal frame (Figure 4-8). The sand bank at the stern area was very steep and eroded with between 50 and 90 cm of 120 cm long fronds exposed across the area. A scour hole was also observed within the hull

of the wreck at the stern. It was noted that the lower parts of the Cegrass™ fronds in this area were clean of marine growth (Figure 4-8) suggesting that they were either newly exposed or are not regularly exposed.

The bow area of the sandbank was in good condition, with no deep scouring seen across the area of Cegrass™ matting, and only between 30 to 60 cm of 150 cm long fronds exposed. A deep scour was seen around the information plinth which lies beyond the Cegrass™ matting at the bow. Parallel lines of seaweed growth in the sand indicate that the hull timbers on the port side in the bow area may have been previously exposed for a time, but had become reburied.

A second area of pebble armour was noted on the starboard side of the wreck towards the bow, in a gap between Cegrass™ mats. The Cegrass™ on the port and starboard side is 90 cm long and exposure of the fronds varied between 30 and 90 cm, with some sections of the metal framing visible. As with the fully exposed fronds at the stern, the lower parts of the fronds showed no evidence of marine growth, again suggesting recent or irregular exposure.

The central area of the wreck, within the hull, also seemed to have changed, with very few barrels visible compared to pre-2006 visits and increased sand cover. The only area of exposed barrels was in the stern area, whilst site plans from the early 1990s show barrels exposed across the entire site. Undulating sand waves were noted across the central part of the wreck, with higher levels on the port side. A ridge of sand was noted just within the port hull at the bow, whilst the starboard side at the bow had areas of exposed timbers both with and without marine growth.

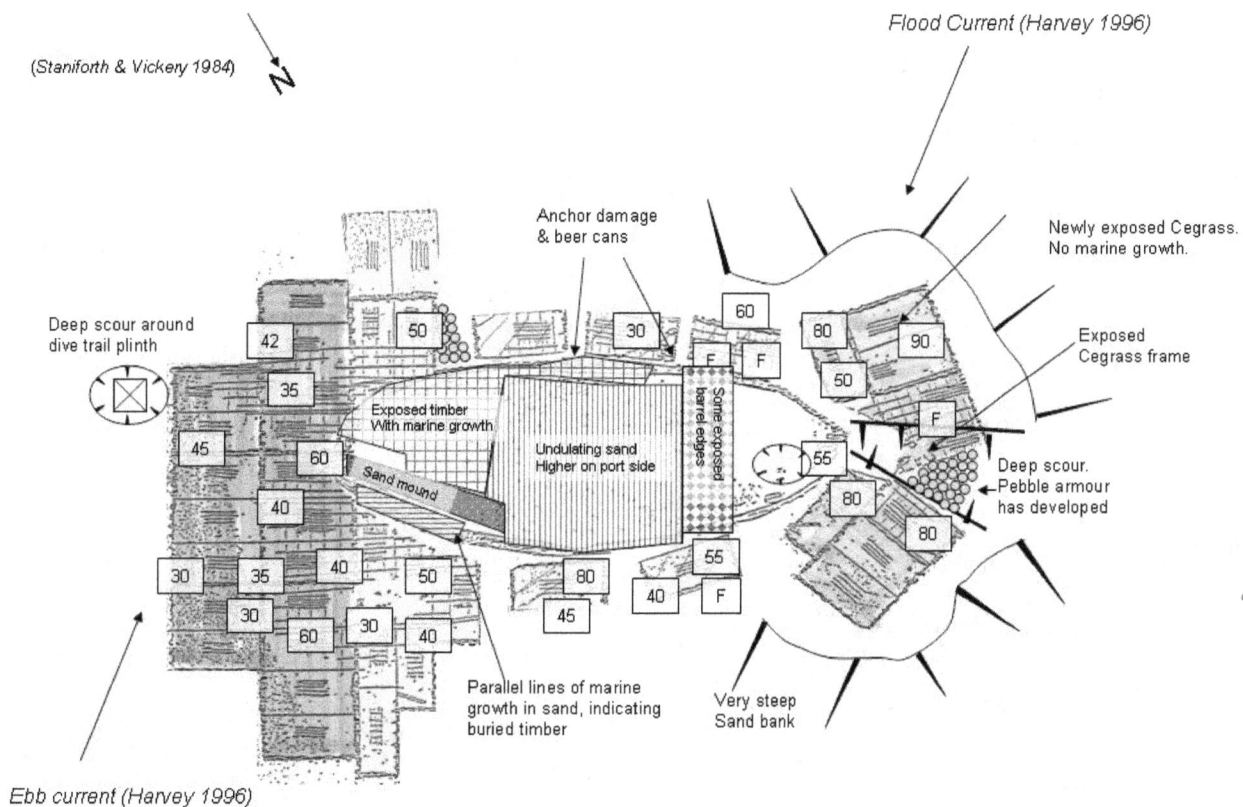

Figure 4-7 Results of 2008 monitoring survey (Heritage Victoria 2008).

A group of recreational divers visited the site at the beginning of February 2009 and found the site to be covered with a thick layer of sand, with very little of the shipwreck timbers visible and the divers were unable to spot any of the barrels. Video footage and photographs from this dive revealed that some of the more obvious and exposed parts of the wreck could be seen beneath a heavy covering of marine growth. Figure 4-9 shows that whilst there is not quite as much sand cover as in 2006, there was certainly more sand than in September 2008.

Discussion

The condition of the site as seen in 2008 was surprising to Heritage Victoria divers who had only seen the site post-Cegrass™ placement, as all previous visits had shown the site to be well supported by the sandbank, with only the tops of the Cegrass™ fronds visible. Heritage Victoria archaeologist Peter Harvey (2008, pers. com.) recalled that when work started on the site and visits were on a monthly rather than annual basis, huge changes were regularly seen in the deposition of sand around the site.

The 2008 inspection posed a couple of questions regarding the current condition and future stability of the site.

1. Is the increased and deeper erosion seen at the stern in 2008 a permanent change to the cycle of erosion and deposition, or within the parameters known on the site in the 1980s and 1990s?

47

2. Is the small number of visible barrels due to increased sand deposition, or due to erosion and loss of the previously visible barrels?

While neither of these questions can be answered without additional regular monitoring of the site, the following observations are relevant.

Figure 4-8 Cegrass™ and metal base exposed during 2008 visit (Heritage Victoria 2008).

Site inspections between 2000 and 2006 were normally undertaken during the summer months, whilst the 2008 inspection was undertaken in September. The sand cover reported during the February 2009 visit suggests that, generally, there is more sand coverage on the site during the summer than during the winter months. What can not be answered without additional dives, however, is whether the deep erosion to the full depth of the Cegrass™ is a regular occurrence, or whether this is a change in the erosion/burial cycle, as is suggested by the lack of marine growth on the lower part of the fronds. It is also unclear as to whether the sand cover will continue to increase during the late summer/autumn of 2009 or whether there has been net sand loss from the site.

The second observation is the differentiation between sand deposition and erosion between the port and starboard sides of the wreck. This might be explained by the prevailing currents which run across the site. The implication being that, during winter 2008 (as seen in September), more sand was being eroded from the starboard side stern area during the flood tide and deposited in the port side bow area than is being replaced in the reverse direction during the ebb tide. As the February 2009 dive was undertaken by recreational divers (for pleasure), no detailed observations regarding sand cover were made, or were visible from the video

Figure 4-9 Partially exposed Cegrass™ fronds in February 2009 (Fuschburger 2009).

footage or photographs.

It is interesting to note that during the first half of 2008, work was completed on deepening areas of the channel as part of the Rip and South Channel as part of a channel deepening works throughout Port Phillip Bay. This work involved the removal of solid substrate at the Rip deepening entrance to Port Phillip. Hydrodynamic and sediment transport models undertaken as part of the Environmental Effects Statements suggested that the Channel Deepening work would not have any effect on the overall sediment transport process in the area of the *William Salthouse* wreck (Port of Melbourne Corporation 2004:28-13). The modelling suggested only small changes in the tidal flows for the area around the *William Salthouse* wreck with a predicted change in tidal current speed of 0.02 m per second at peak ebb-tidal flow (Port of Melbourne Corporation 2004:28-4). Monitoring works on sand levels and tidal flow changes in the area have not been published since the Channel Deepening project has progressed, and it is unclear at this stage whether the works have had any effect on the stability of the *William Salthouse* wreck.

Conclusion

It is clear that sandbagging and Cegrass™ matting have been an outstanding success as an example of *in situ* stabilisation of an intact wooden shipwreck, preserving a large section of hull and cargo for over ten years. The combination of diver education and the permit-only diver access program has also served the site well, by maintaining interest amongst Victorian divers, yet also protecting the site from diver interference and inadvertent damage.

There are clearly a number of issues which must be addressed in order for the site to survive into the future, namely illegal access into the protected zone and the recent sand loss. Whilst policing shipwreck sites in Port Phillip Bay is difficult, Heritage Victoria continues to build relationships with the water police, and to work on wider advertising of the protected zones amongst the fishing community. The scouring observed in 2008 needs addressing, and a program of monitoring is planned. An evaluation of any changes to the hydrodynamics and sediment transport systems in the area needs to be assessed, and it may be necessary to place new Cegrass™ sheets around the stern to stabilise this area. If it appears that previously preserved cargo items are being eroded and lost in the tides, then it may be necessary to investigate new measures to prevent further deterioration. Although this would compromise the attractiveness of the site to sport diving visitors.

Acknowledgements

Thanks must go to the 2008 dive team Peter Harvey, Cassandra Philippou, Rhonda Steel, Sven Bartels and Isa Loo who were all outstanding in assisting with the data collection for this paper, and to Peter particularly for his help with the early years. I would also like to thank Martina Fuschberger who passed on video footage and photographs of her visit in February 2009.

References

Harvey, P.
1996   A Review of Stabilisation Works on the Wreck of the *William Salthouse* in Port Phillip Bay, *Bulletin of the Australian Institute for Maritime Archaeology* 20(2):1-8.
Hosty, K.
1989   Bagging the *William Salthouse*: Site Stabilization Work on the *William Salthouse*, *Bulletin of the Australian Institute for Maritime Archaeology* 12(2): 13-16.
ICOMOS (Australia)
1981   *The Australia ICOMOS Charter for the Conservation of Places of Cultural Significance (The Burra Charter)*, Sydney.

Port of Melbourne Corporation.
2004   Channel   Deepening   Project,   *Environmental   Effects   Statement*, http://www.channelproject.com/publications/effectsstatement/ees_technical.asp Accessed 8 May 2009

Staniforth, M. and L. Vickery
1984   *The Test Excavation of the* William Salthouse *Wreck Site, An interim report*, Australian Institute for Maritime Archaeology, Special Publication No. 3.

Strachan, S.
1988   *The* William Salthouse *(1824-1841) Conservation Plan*, Manuscript, Victoria Archaeological Survey, Ministry for Planning and Environment, Melbourne, Victoria.

5   Corrosion Monitoring and the Environmental Impact of Decommissioned Naval Vessels as Artificial Reefs

Vicki Richards
Department of Materials Conservation, Western Australian Museum, Shipwreck Galleries, 45-47 Cliff Street, Fremantle, Western Australia, 6160

Ian MacLeod
Collection Management and Conservation, Western Australian Museum, Locked bag 49, Welshpool DC, Western Australia, 6986

Peter Morrison
Sinclair Knight Merz Pty Ltd, PO Box H615, Perth, Western Australia, 6001

The former Australian Naval vessels, HMAS *Swan* and HMAS *Perth* were scuttled as artificial reefs and recreational dive sites off the Western Australian coast in 1997 and 2001, respectively. In addition, the former HMAS *Hobart* was sunk in Gulf St Vincent, South Australia in 2002. During the preparation of the ships for sinking, significant quantities of metals, such as copper alloys, aluminium, lead and steel, and petroleum hydrocarbons and other potential pollutants were removed, however, there remains the possibility that the corrosion of the submerged vessels and the presence of residual hydrocarbons may impact on the local marine environment. Hence, corrosion monitoring programmes were implemented for the three vessels and concentrations of key contaminants, such as heavy metals, total petroleum hydrocarbons and butyl tins in the surrounding sediments were monitored periodically. The results provided important information on the synergistic interactions between modern shipwreck materials and the marine environment.

Introduction

Artificial reefs are not a new concept. For centuries Japanese fishermen have attempted to increase the fishing productivity of their local waters by dumping rocks into the sea (Russel 1975). During the 1960s and 1970s over 100 artificial reefs were constructed in the coastal waters off the USA and many other countries including Australia. In the 1970s, several tyre reefs were constructed in Gulf St Vincent and Spencer Gulf, SA and derelict ships and barges were sunk off the coast of Sydney, NSW. Similarly, three vessels were scuttled and three artificial tyre reefs were established off the WA coast in an attempt to enhance recreational fishing and diving (Morrison 2003). The colonisation and/or corrosion of these artificial reefs were not monitored rigorously and there is little, if any data

published for comparative studies. The scuttling of decommissioned Royal Australian Navy (RAN) vessels as artificial reefs has provided the opportunity to study the environmental impact of these enormous contemporary ships on the local marine environment and obtain biological and corrosion data from the time of sinking. This latter data is important as this type of information is usually gathered from historic shipwrecks that have been submerged for a considerable period of time and not from the initial stages of wrecking. The results should provide important information on the long-term stability of the vessels and the synergistic interactions between modern shipwreck materials and the marine environment. This knowledge is becoming more important as it is apparent that more of these decommissioned naval vessels and confiscated illegal fishing boats will be sunk as artificial reefs in the near future. Perhaps more importantly, the information can be used to better understand the deterioration of historic shipwreck sites and ultimately assist in the development of appropriate *in situ* management strategies for underwater cultural heritage sites.

The Vessels

Swan (DE50) was an Australian built River class destroyer escort, commissioned in 1970. It was 113 m in length and 24 m tall from keel to the tower, with a beam of 12.5 m and a 5.3 m draught. *Perth* (DDG38) and *Hobart* (DDG39) were American built Charles F. Adams class guided-missile destroyers, commissioned in 1965. They were approximately 134 m in length and 38 m tall from keel to tower, with a beam of 14.2 m and a draught of 6.1 m (Gillet and Graham 1977; Chant 1984; Gillet 1986, 1988; Odgers 1989). In order to interpret the environmental and corrosion data accurately it is necessary to have some knowledge of their major metal compositions and the paint formulations used on the vessels. The hulls were primarily mild steel plate and possibly small quantities of a carbon-manganese steel. The superstructure was aluminium alloy, probably 5083 (4.5% magnesium) primarily used for welded plate structures but thin material, such as furniture, ductwork, panel linings, awnings and general sheet metal work could have been 5052 (2.2% magnesium). In the late 1980s and 1990s the RAN had been using organotin self-polishing paint so the paint system below the waterline would have been a vinyl anti-corrosive plus antifouling International Intersmooth Self Polishing Co-polymer, which contained tributyl tin and cuprous oxide. The original paint above the waterline was probably zinc chromate primer with alkyd enamel undercoat and topcoat. Some or all of this may have been replaced in the 1980s and 1990s with zinc

rich epoxy primer, epoxy intermediate coats and alkyd topcoat.

All hydrocarbons, hazardous material and liquids, debris and, as far as was practicable, all metals of environmental concern were removed prior to scuttling in accordance with the Environmental Management Plans (EMPs). This was undertaken to minimise the environmental impact as a result of heavy metal and petroleum hydrocarbon contamination. During the preparation process, a number of measuring points were attached to predetermined positions on the hull and superstructure of the vessels to facilitate the corrosion monitoring programmes.

After the preparations were completed and the vessels had passed rigorous inspections by Environment Australia, *Swan* was scuttled on 14 December 1997 in Geographe Bay, Dunsborough, WA, *Perth* on 24 November 2001 in King George Sound, Albany, WA and *Hobart* on 5 November 2002 in Yankalilla Bay, Gulf St Vincent, SA. The post scuttling monitoring regimes for each vessel varied; however, they included a combination of sediment sampling for detection of heavy metals, total petroleum hydrocarbons and butyl tin, ecotoxicological assessments, recording the rate of colonization by fishes and encrusting marine life and corrosion surveys.

Experimental

Monitoring Programmes

The monitoring schedules for the corrosion and sediment surveys of the vessels are summarised in Table 1 at the end of article. The monitoring programme for *Swan* required biological and sediment surveys in accordance with the Sea Dumping Act (1981) (Environment Australia 1984). The fish community and the encrusting biota on the wreck were monitored regularly over the first five years and the results of these surveys were reported in Morrison (2003). Marine sediment samples were collected from both the control site and the proposed scuttling site for *Swan* during the baseline survey and at intervals of five and twelve months after scuttling. The sediments at the wreck site and the selected reference site were analysed for a suite of heavy metals [aluminium (Al), iron (Fe), cadmium (Cd), chromium (Cr), copper (Cu), lead (Pb) and zinc (Zn)] and total petroleum hydrocarbons (TPH).

The sediment sampling programme for the *Perth* site was in accordance with the EMP governed by the Western Australian Environmental Protection Act under the supervision of the Department of Environmental Protection (DEP) (Morrison 2007). The post-scuttling surveys were undertaken after six months, one, three and five years. The sediment was analysed for nickel (Ni), tin (Sn) and mercury (Hg) in addition to the aforementioned heavy metals, TPH, total organic carbon (TOC) and tributyltin (TBT). Ecotoxicological assessment of the sediment during the baseline sediment survey was also performed in order to separate the ecological effects of the vessel from any existing contaminants; however, because the levels of TBT and metals in the sediments did not exceed ANZECC/ARMCANZ (2000) guidelines over the five year survey, the toxicity testing was not carried out after the initial baseline assessment and hence, the ecotoxicological results will not be presented in this paper. The rate of colonisation of *Perth* by fishes and encrusting marine organisms was not recorded. The results of the five-year biological survey of *Swan* could be extrapolated to this site so it was deemed unnecessary.

A baseline benthic survey of the proposed wreck site for *Hobart* prescribed by the Environmental Protection Authority (Morrison 2002) was conducted to determine the existing assemblage and

sediment quality. Unfortunately, no further sediment or biological monitoring has been undertaken at this site; therefore the results of this baseline survey will not be presented.

Corrosion Monitoring

The corrosion surveys involved measuring the corrosion parameters of a number of stainless steel 316 (SS) bolts by well trained diving pairs. The bolts were attached to various positions on the steel hull and the aluminium superstructure of *Swan*, *Perth* and *Hobart* prior to scuttling. The attachment and documentation procedures are outlined in Richards (2003) and the corrosion parameter measuring procedures (corrosion potential (E_{corr}) of the SS bolt and surface pH of the adjacent metal surface) are reported in MacLeod et al. (2004). The water depth at each position was measured with a digital dive computer.

Sediment Monitoring

Replicate sediment samples from the *Swan* scuttling and reference sites (3 km north west of the scuttling site) were collected by divers from random locations along a 100 m north-east transect using a polycarbonate hand corer inserted to a depth of 2 cm. Replicate sediment samples from the *Perth* scuttling and reference sites (1.7 km north of the scuttling site) were obtained at the prescribed locations using a Van Veen grab and then sub-sampled for analysis. Samples from the *Perth* site were collected at distance intervals of 10 m, 50 m 125 m and 500 m away from the vessel, along two perpendicular axes (south and west). The detailed sediment sampling and analytical procedures for *Swan* are described fully in Morrison (1998) and for *Perth* in Morrison (2007).

Results

Corrosion Monitoring Programme

The measuring point positions are shown diagrammatically on the general arrangement plans for *Swan* and *Perth* in Figure 5-1. The measuring point positions, 19 and 20 on *Perth* were inadvertently cut out some time after attachment, prior to scuttling. The measuring point positions on *Hobart* are almost identical to those of *Perth* (Figure 5-2); however, the Mount 51 gun on the main deck towards the bow of *Hobart* remains intact. The corrosion potential and surface pH of each point measured on *Swan, Perth* and *Hobart* at the specified time intervals are shown in Tables 2, 3 and 4 at the end of article. The surface pH of the metal surfaces adjacent to the bolts were not measured until the vessels had been submerged for at least one year to allow the ships to attain some form of 'steady state' with the local marine environment.

Sediment Monitoring Programme

The sediment survey schedules are outlined in Table 1 and in accordance with the individual EMPs for each vessel, no further sediment monitoring is required on any site. The detailed analyte concentration data for every sediment survey performed over the full monitoring period specified for each vessel are published in the final reports by Morrison (1998) for *Swan*, Morrison (2002) for *Hobart* and Morrison (2007) for *Perth* and will not be reproduced in this paper; however, the interpretation of the data will be discussed.

Discussion

Corrosion Surveys

The main objectives of the corrosion monitoring programmes are to measure the corrosion parameters of previously designated sites on each of the vessels at specified time intervals, to ascertain any discernible differences in the corrosion rates between the various locations on the ships and to ultimately monitor the long-term stability of the vessels. The results of the corrosion monitoring programmes will also assist in interpretation of the sediment survey data with respect to any

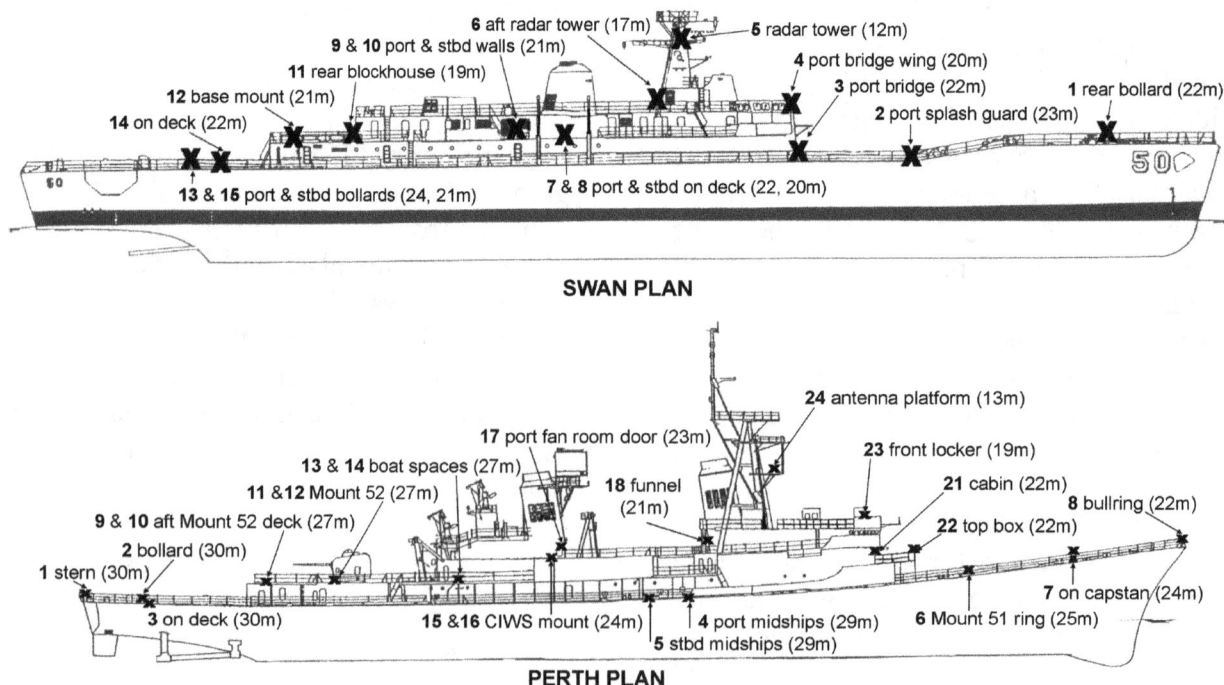

Figure 5-1 General arrangement plan of *Swan* and *Perth* indicating the position of the 15 and 22 measuring points respectively (Author).

contamination by heavy metals.

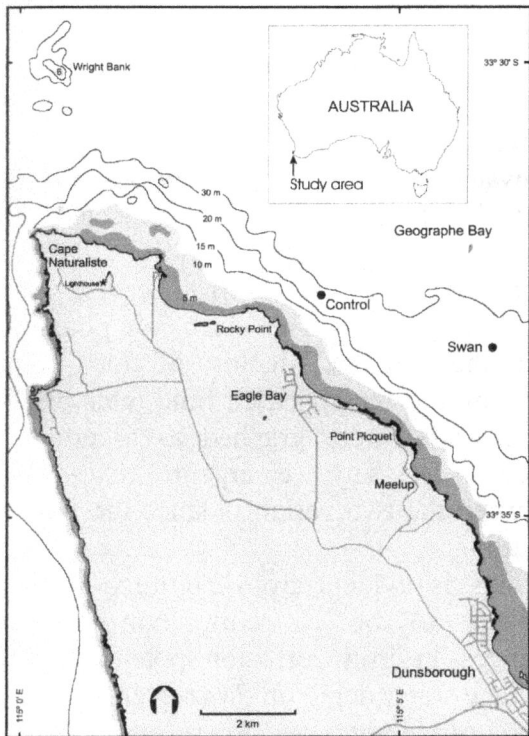

Figure 5-3 Location of *Swan* including the sediment sampling positions (Morrison 2003:6).

The wreck of the *Swan* lies 2.4 km north-east of Point Piquet, Dunsborough in Geographe Bay, Western Australia (Figure 5-2) in 31 m of water at high tide whilst the tower reaches to within 8 m of the water surface. The vessel rests on the keel, the bow facing north-west with a 10 degrees list to port (MacLeod et al. 2004). *Perth* lies in King George Sound, Frenchman Bay, Albany, Western Australia about 9.5 km south-east of Albany (Figure 5-3). The vessel rests on the keel with the bow facing approximately east with no noticeable list to either side. The total depth to the keel at the sediment line is about 34 m. The remains of the radar tower rises 4 m out of the water surface to act as a navigation marker (Richards and MacLeod 2004). The authors warned that unless the attachment points of the radar tower were substantially

strengthened prior to scuttling then the tower would eventually fail due to differential aeration corrosion. This occurred in 2003, after only two years of exposure but the tower was subsequently reattached with stainless steel guy wires. *Hobart* lies

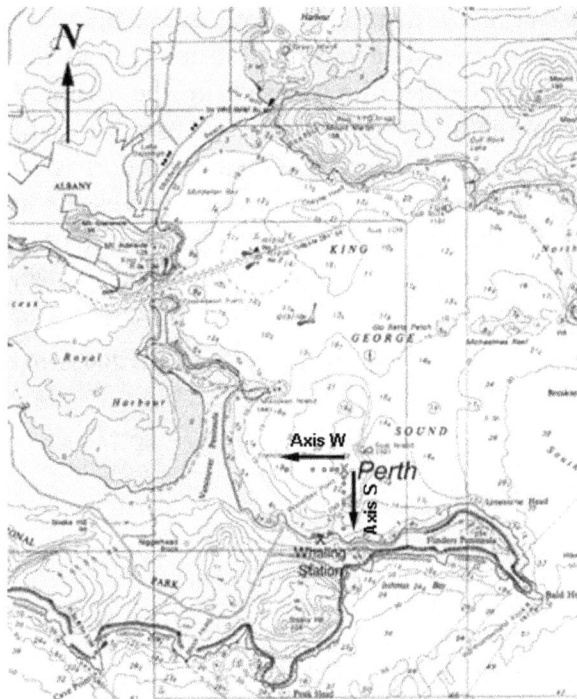

Figure 5-2 Location of *Perth* including the sediment sampling positions (MacLeod et al. 2004:61).

offshore, about 7.5 km west of Marina St Vincent, Wirrina Cove in Yankalilla Bay, Gulf St Vincent, South Australia (Figure 5-4). It lies in a roughly east-west orientation with no noticeable list to either side, with the bow facing approximately east. The total depth to the keel at the sediment line is about 28 m, the depth to the main deck is 23 m, while the remains of the radar tower rises to within 5-6 m of the water surface dependent on the tidal range (Richards 2003).

Corrosion monitoring of the vessels was facilitated by measuring the corrosion potentials of stainless steel bolts that had been attached to the hull and superstructure, prior to scuttling. This provided high profile,

54

long-term electrical ohmic contact with the vessels, ensuring that the same positions were measured over the course of the monitoring programmes. This consistency ensures that comparisons of any differences and/or similarities in corrosion behaviour between points on the same vessel and between the different vessels are as accurate as possible. The physical structure and the nature of the alloys used in the original manufacture of these vessels will significantly affect the measured corrosion potentials. The orientation of the vessel on the seabed in relation to the physical (e.g. water movement and depth, tidal range, current direction and speed, etc.) and

Figure 5-4 Location of *Hobart* (Richards 2003a:6)

chemical (e.g. dissolved oxygen concentration, salinity, temperature, etc.) conditions of the surrounding marine environment will also have a significant affect on the corrosion results.

The total amount of water movement generally decreases with increasing water depth thereby reducing the total amount of oxygen flux to a metal surface. The general rate of deterioration of concreted metals on shipwreck sites is very dependent on the water depth and the flux of oxygenated seawater over the objects lying proud of the seabed. Therefore it is expected for concreted iron alloy artefacts that the corrosion potentials will decrease with increasing depth. The corrosion potentials and the average depths of the points measured at the specified time intervals for each vessel were graphed to ascertain if there was any emergent relationship between the two variables since the time of scuttling.

As is evident from Figure 5-5, there appears to be no correlation between changes in iron corrosion potentials with average water depth on *Swan* even after four years of submersion. There was also no emergent trend between *Swan* aluminium alloy voltages versus water depth. Similarly, the corrosion potentials for the iron and aluminium alloys on *Perth* and *Hobart* showed no correlation with increasing water depth. The lack of any apparent dependence of E_{corr} on depth is simply a reflection of the inter-connected nature of the structural elements of the vessel and that large sections of the ships are still electrically connected with each other; therefore, the voltages expressed do not relate directly to the measurement point. The ships are still very much in the initial stages of deterioration and although there has been some colonisation of the vessel's surfaces, due to the anti-foul present below the waterline of the vessel and the protective paint systems, no significant concretion formation has occurred on the steel hulls during their short period of immersion. Therefore, the usual film free iron corrosion mechanism applicable to concreted iron objects where the anodic and cathodic sites are separated by a semi-permeable concretion membrane does not apply to these vessels at this point in time. Experience with the wreck of *Fujikawa Maru* (1944) sunk in Chuuk Lagoon, Federated States of Micronesia during WWII has shown that this wreck is only now just beginning to have E_{corr} values that are sensitive to water depth (MacLeod 2003). *Fujikawa Maru* has a very similar site orientation to *Perth* in that it is lying upright on its keel on a flat seabed so it is expected that a decade or two will be needed before any E_{corr} values on the naval vessels show any systematic behaviour with regard to water depth.

It is vital that the data from the modern naval vessels are viewed in their own right since the fundamental corrosion microenvironment is different to that of historical concreted marine iron that has been immersed for more than 100 years. The different processes mean that the rate-determining step in the overall corrosion process at this stage is not the reduction of oxygen on the surface but the oxidation of the metal. All three vessels are painted with an anti-corrosive paint system, which will

significantly reduce the corrosion rate of the underlying metals, decrease concretion formation and subsequently, affect the primary corrosion mechanisms. In an attempt to understand changes in the corrosion potentials with time, the voltages of the aluminium and iron alloy points on each vessel were plotted against the time of submersion since the scuttling.

Since the corrosion behaviour of *Perth* and *Hobart* are very similar, only *Perth* results will be presented and the concomitant interpretation of that data can be extrapolated to the corrosion behaviour of *Hobart*. The changes in the iron and aluminium alloy corrosion potentials of *Perth* over the three years of submersion are shown graphically in Figures 5-7 and 5-8, respectively. The baseline corrosion potential of point 7 (-0.434V) on *Perth* appears to be anomalous and may have been caused by partial electrical connection between the platinum electrode and the bolt during measurement. There is a relatively random scatter of points during the first six months (0.5 year) of immersion for both the iron alloy (Figure 5-6) and aluminium alloy

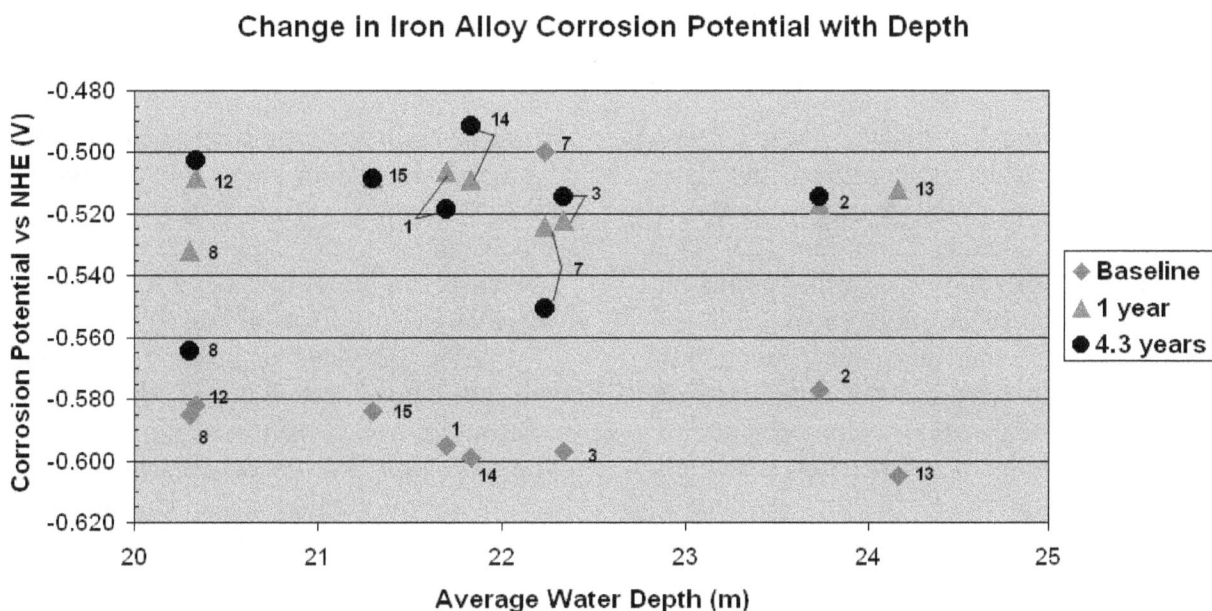

Figure 5-5 The relationship between the iron alloy corrosion potentials and depths for the *Swan* (Author).

points (Figure 5-7). This random scatter may reflect differences in the amount of sea water penetration through and under the paint film in the vicinity of the measuring points and this, in turn, would lead to a range of dry to wet corrosion cell mechanisms occurring under the paint layers. Therefore, interpretation of any differences between individual data points or sets of the same is not possible for the first six months of immersion.

cathodic sites, effectively slowing the corrosion rate. Therefore, it is more likely that this decrease in the average corrosion potentials coupled with slightly more acidic pHs at the metal surfaces after sixteen months (Fe pH = 8.09 ± 0.13; Al pH = 8.12 ± 0.07) than the surrounding seawater (average pH = 8.26) is more consistent with a slight increase in the corrosion rate due to the development of localised corrosion cells on the vessel as the seawater penetrates the

Figure 5-6 The change in iron alloy corrosion potentials on *Perth* with time (Author).

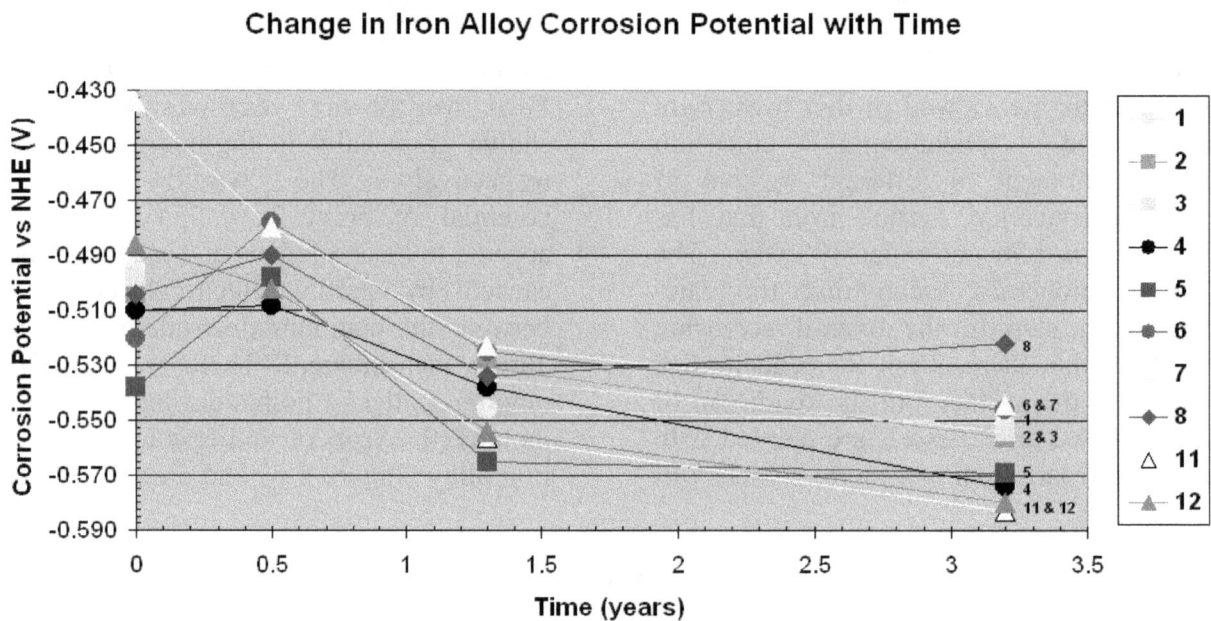

After sixteen months (1.3 years) there was a small, relatively consistent decrease (more negative) in the average corrosion potentials of both the aluminium (0.054V) and iron (0.047V) alloy measuring points on *Perth*, reflecting more effective and uniform penetration of the seawater under the protective paint film and the establishment of a series of localised corrosion cells after a further ten months of submersion. Although there is an increase in the secondary colonisation of the hull and superstructure after this time there is no significant uniform formation of concretion, which would cause some partial separation of the anodic and

paint layers, causing pitting corrosion of the iron and aluminium under this passive paint film. Unlike film free corrosion mechanisms, with pitting corrosion a decrease in corrosion potential indicates an increase in the corrosion rate.

After 3.2 years (2.2 years for *Hobart*) the corrosion potentials of the iron and aluminium points have essentially stabilised (i.e. 1.3 and 3.2 years average corrosion potentials are within the standard deviations) indicating that the vessels have attained some form of 'steady state' with the local environment. However, small differences in the metal corrosion potentials are becoming apparent and the surface pH measurements of many of the iron and aluminium points have decreased significantly (Fe pH = 7.85 ± 0.56; Al pH = 7.70 ± 0.80), which may indicate some slow changes in corrosion mechanisms occurring on the vessel. It is possible that the increased time of immersion could have caused some partial separation of the anodic and cathodic sites on some points effecting small changes in the corrosion rates. Under these circumstances, a decrease in the corrosion potentials (more negative voltages) could indicate a relatively small decrease in the corrosion rate of the vessel.

The more positive potentials of the iron alloys (Figure 5-6) measured on the bow (8 – bullring; 6 – Mount 51 ring; 7 – capstan) and the stern (1 – stern; 2 – bollard; 3 – on deck) suggests that these areas are subjected to a more aggressive environment than the more protected points located midships on the port (4) and starboard (5) side of the lower deck and the Mount 52 gun on level 1 (11 – port side; 12 – starboard side). Due to the orientation of the vessels on the seabed and the physical structure of the ships themselves, it would be expected that there would be more local turbulence and water movement occurring around the bow and stern areas, increasing the overall oxygen flux to these metal surfaces and in turn, increasing corrosion rates.

The aluminium alloy corrosion potentials seem to have stabilised after 3.2 years (Figure 5-7) but the differences between the voltages are not as pronounced as those noted for the iron alloy points. Although decreases in the surface pH of some of the points indicate that there have been some breaches in the paint film, the smaller potential differences suggest that this protective layer remains relatively intact, and therefore, the aluminium superstructure is less affected by changes in the local environment than the steel hull structure over this same time period. In addition, aluminium is less prone to concretion build-up and the corrosion mechanism is very different to that of partially concreted iron; therefore, any changes in the aluminium corrosion rates

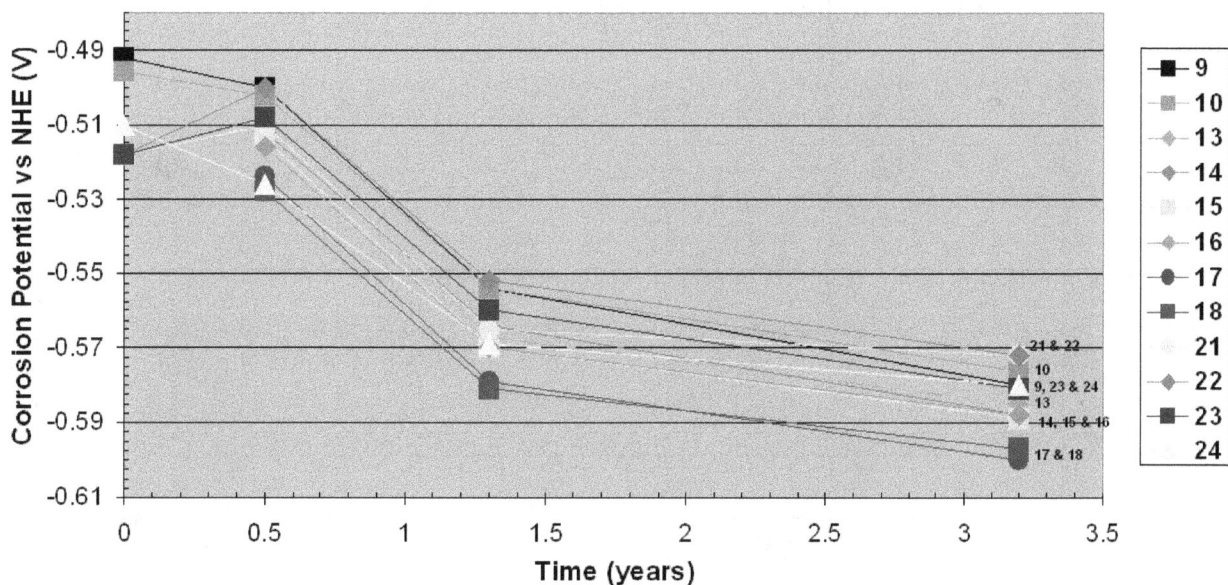

Figure 5-7 The change in aluminium alloy corrosion potentials on *Perth* with time (Author).

would be significantly slower than on the ferrous surfaces.

The changes in the iron and aluminium alloy corrosion potentials of *Swan* over the four years of submersion are shown graphically in Figures 5-8 and 5-9, respectively. The baseline iron and aluminium corrosion potentials measured fours hours after scuttling are indicative of dry cell corrosion mechanisms as the seawater would not have had time to effectively penetrate the protective paint system. Again the voltage of point 7 (port side on deck 1) appears anomalous. After one year, the voltages become more positive, similar to the voltages measured on *Perth* after sixteen months, reflecting increased deterioration and penetration of the paint barrier with seawater causing a corresponding increase in the formation of localised wet corrosion cells. Again, after four years the iron corrosion potentials have stabilised and the vessel appears to have attained some form of 'steady state' with the local environment. However, the differences in the iron and aluminium corrosion potentials are more pronounced than those measured on *Perth* and *Hobart*, which may indicate more significant changes occurring on *Swan* in comparison to the other vessels.

The more positive voltages of the iron alloys measured on the stern (Figure 5-8) (12 – upper deck on mount; 13 – port bollard; 14 – on deck forward of starboard bollard; 15 - starboard bollard) and the bow (1 – rear bollard; 2 – port splash guard; 3 – port side below bridge) suggests that these areas are subjected to a more corrosive environment than the more protected points (7 – port side; 8 – starboard side) located flush on the upper deck, midships on the vessel. There is a significant amount of scouring around the stern and bow of the vessel, which indicates that considerable water movement is occurring in these areas, subsequently increasing the corrosion rates of these bow and stern points. The voltage of point 13 (stern bollard) at –0.368V is very close to the open circuit potential for iron in flowing seawater (-0.383V). This voltage suggests that the bollard is not in good electrical contact with the rest of the hull. This result supports the fact that isolated iron artefacts corrode at a faster rate than a

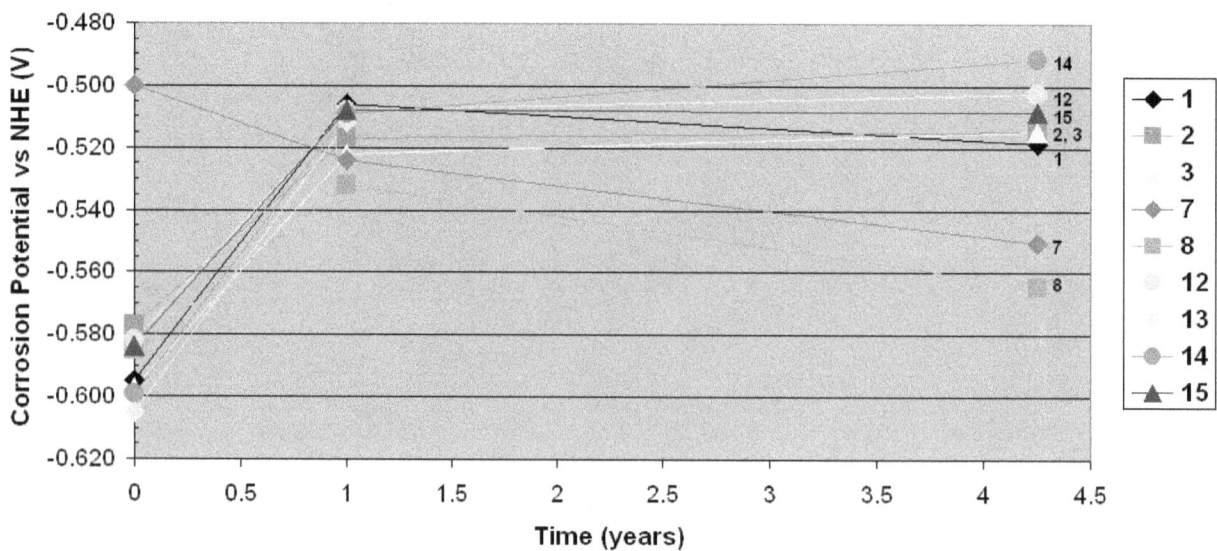

Change in Iron Alloy Corrosion Potential with Time

Figure 5-8 The changes in iron alloy corrosion potentials on *Swan* with time (Author).

59

large, intact iron structure. This is due to the massive difference in surface area and therefore, the current density is dispersed over a larger area and the corrosion rate of the hull will significantly decrease in comparison to the isolated iron fitting.

The aluminium alloy points (Figure 5-9) are reacting similarly to the iron alloys with respect to total water movement and localised turbulence factors. The less negative corrosion potentials of aluminium points 5 (radar tower), 10 (upper deck, port side on wall) and 11 (rear blockhouse on wall) indicate they are subject to a more corrosive environment than positions 4, 6 and 9. Point 4 is positioned on the port side corner of the bridge behind a large splash guard, point 6 is mounted on the wall on the rear of the radar tower and point 9 is mounted on the wall, behind a ladder on the starboard side of the superstructure. All three points are considerably more protected from excessive water movement than the other more exposed positions.

Galvanic corrosion may also be contributing to the emerging differences in the iron and aluminium alloy corrosion potentials on *Swan*. The effects of galvanic corrosion are most readily discerned when the iron alloy hull and the aluminium of the superstructure are in direct electrical contact. When this occurs the aluminium will act as an enormous sacrificial anode for the steel hull promoting galvanic corrosion of the aluminium and protecting the steel structure. Naval architects make enormous efforts to ensure electrical isolation of the different alloys used in the construction of these vessels to prevent galvanic corrosion occurring during service so it will be only after this insulation has begun to fail and the seawater has penetrated the isolation barriers that galvanic corrosion will be observed. The anode or cathode resistances in the galvanic cells are controlled by protective oxide films and the resistance associated with the custom-made industrial paint system.

Change in Aluminium Alloy Corrosion Potential with Time

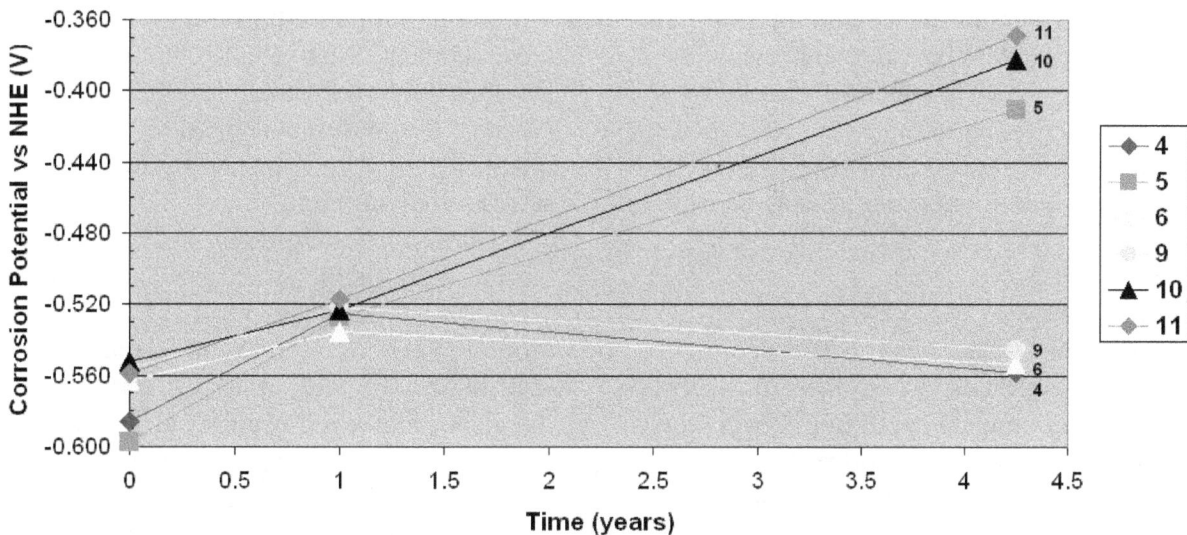

Figure 5-9 The changes in aluminium corrosion potentials on *Swan* with time (Author).

60

Data from *Swan* after four years indicate that breaches in the electrical barriers have commenced but they still appear to be intact after 2 and 3 years of immersion for *Hobart* and *Perth*, respectively. On *Swan*, points 7 and 8 (midships, flush with deck 1) appear to be receiving some cathodic protection by the aluminium superstructure, while the stern and bow are the least protected (more positive voltages). Preferential corrosion of the aluminium is also indicated by the decrease in the average pH of the aluminium elements (8.03 ± 0.24) and an increase in the average surface pH of the iron measurement points (8.29 ± 0.40).

There were a significant number of discrete areas of aluminium corrosion products visible on the planar surfaces of each vessel's aluminium superstructure, which increased significantly with the time of immersion. Pitting and crevice corrosion are causing this deterioration of the aluminium as the protective paint system is slowly failing in these areas but it has not significantly affected the overall corrosion rates even after four years of submersion. These results signify that these contemporary vessels are corroding albeit at a relatively slow rate and this is most probably due to the protection afforded by the paint barrier remaining on the vessels and the insulating techniques used to prevent galvanic corrosion when the vessels were in service.

Sediment Surveys

The main objectives of the sediment analysis programmes are to ascertain any environmental impact resulting from the scuttling of these contemporary vessels, to ascertain the extent of metal enrichment of the sediments surrounding the submerged vessel, to ascertain if any change in corrosion mechanism has an impact on the extent of metal enrichment in the sediment and to ultimately monitor the long-term

stability of the vessels. During the decommissioning of the ex-naval vessels significant quantities of metals and oil were removed as scrap for both the purposes of salvage and for reducing the potential environmental impact. Vast quantities of copper, brass, aluminium, lead and steel were removed; however, it was not possible to remove all traces of metals, especially copper and brass from the engine rooms. The metals remaining on the vessels will corrode and diffuse into the surrounding environment, therefore it is imperative that the sediments in close proximity to these vessels are monitored for these major contaminants and the environmental impact assessed.

Prior to the scuttling of *Swan*, chromium and iron were found to be significantly elevated indicating that the site had naturally elevated levels of these two meals compared with the reference site. No statistically significant differences in sediment metal concentrations between the reference and *Swan* site were noted after five months. After a year there was a marked enrichment of all heavy metal analytes (Al, Cr, Cu, Fe, Pb and Zn) in the sediments directly adjacent to the vessel but only copper was found to exceed the Environment Australia (2002) guidelines. None of the sediments contained measurable quantities of total petroleum hydrocarbons after twelve months.

Metal enrichment of sediments surrounding metal structures and jetties is common. Since the area of enrichment around such structures is dependent on distribution by currents, the sediments on the *Perth* site were monitored at increasing distances (10, 50, 125, 250 and 500 m) away from the vessel in the major current directions (south and west) for that area over a five year period. Whilst there was some variability in metal levels measured at the reference locations, none were statistically significant. The concentration of cadmium and mercury did not increase significantly above the Practical Quantitation Limit. The concentrations of the other detectable heavy metals (Al, Cr, Cu, Fe, Pb, Ni, Sn and Zn) showed an initial increase since the baseline survey followed by a decline after five years; however none were above the screening level in the Australian and New Zealand guidelines for fresh and marine water quality (ANZECC/ARMCANZ 2000). After five years most of the metals were an order of magnitude less than the guideline screening level for that metal. Typically metal enrichment in the sediment appeared to be localised and restricted to within 50 m of the vessel and there was no significant differences in the heavy metal distribution in the sediment measured along the west and south transects. The tributyltin (TBT) results indicate that there was some contamination of the *Perth* site within 50 m of the vessel after one year; however, the TBT was below screening levels at all locations after five years post-scuttling. It is possible that a paint flake had dislodged from the hull during the scuttling process. The significant variation between replicate samples supports this suggestion; however, after five years this initial elevation had subsided, likely as a result of natural decomposition.

' Metal enrichment of the sediment surrounding *Swan* and *Perth* is a direct result of metal corrosion and the degradation of the protective paint layers. The major source of aluminium and iron in the sediment would originate from the corrosion of the superstructure and the hull, respectively. Aluminium flake is also a constituent in the primer applied to the keel up to the waterline and iron is a minor constituent in the alkyd resin topcoats. Galvanic corrosion of the aluminium in preference to the iron hull would cause significant increases in the levels of aluminium with a corresponding plateau or only very slight increases in the concentration of iron in the sediment.

Hence, the increases in the aluminium and iron levels measured in the sediment over the *Perth*'s five year monitoring period support the corrosion survey results, which indicate that minimal cathodic protection is being afforded by the aluminium superstructure to the steel hull after five years in an aerobic seawater environment.

Chromium and nickel are alloying metals in stainless steel; however, corrosion of stainless steel occurs predominantly under anaerobic conditions and since both vessels lie essentially proud of the seabed in aerobic seawater this would be only a minor source of contamination. More likely the chromium has originated from the original yellow zinc chromate primer that is commonly used on aluminium alloys. Although some or all of the original coatings were replaced in the 1980s and despite recent coating formulations that did not contain chromium salts, the extent to which the original coatings were stripped prior to repainting is unknown. The copper, zinc and tin in the sediment would originate from the corrosion of copper and copper alloy components on the vessel; however, much of this material was removed prior to scuttling. Copper linear flex shaped charge explosives were used to scuttle *Perth* and may have also contributed to the contamination; however, if the explosives were the major source of the contamination, then increases in the sediment copper levels over time would not be expected. It is more probable that the major source of copper and tin would be from the corrosion of residual copper alloys from the engine room and copper oxide and organotin, which are the major constituents of the recent anti-foul.

The anti-foul is a one pack, organotin based antifouling, self polishing copolymer (Intersmooth Hisol) used for vessels larger than 25 m. The coating contains tributyltin and the dry copolymer film contains 2.2% tin. The biologically toxic organotin and copper compounds leach into the water column and prevent marine organisms attaching to the external immersed hull sections. The TBT leaches into the water column and degrades slowly in the marine environment to the tributyl, dibutyl and monobutyl species and finally to inorganic tin. The environmental degradation of TBT is principally biologically mediated and closely follows first order kinetics. The speciation products of organotin degradation are complicated by accumulation and degradation processes, which occur at different rates depending on the environmental compartment but from the results, initial contamination on the *Perth* site has subsided after five years.

The degradation of the original zinc chromate primer could account for some of the zinc enrichment but the major source would be from the zinc salts, such as zinc phosphates and zinc powder added to the more recent primers as corrosion inhibitors. Lead was used as ballast in these vessels but due to its biological toxicity it was all removed during the preparation process prior to the scuttling in accordance with the EMPs. No lead oxides were present in the most recent paint formulations; however, they may have been used in some of the original coatings and could possibly be a source of the lead contamination. During the scuttling of *Swan*, the explosives used were lead linear flex shaped charges, which would have contributed some lead to the surrounding environment and presumably to the seabed sediments. The elevated levels of lead found in sediments near *Swan* probably resulted from the explosive charges. As a result, lead linear flex shaped charges are no longer used for scuttling vessels.

The sediment results indicate that after five years of immersion in an aerobic marine environment the metal components on *Perth* and *Swan* are corroding but minimal cathodic protection is being afforded to the vessels and the protective paint coatings are deteriorating albeit at a relatively slow rate.

Overall, the results indicate that the scuttling of *Swan* and *Perth* has had no adverse environmental impacts on sediments and it is unlikely the enrichment at these levels will impact significantly on marine life.

Conclusions

The monitoring programmes for *Perth*, *Swan* and *Hobart* are the first systematic corrosion survey programmes established in Australia for monitoring the long-term stability of twentieth century warships after scuttling as artificial reefs. The results of the corrosion and sediment monitoring programmes have shown that the corrosion behaviour of the three vessels are very similar over the first twelve months of immersion and corrosion rates increase over time, especially after the first year. However, the vessels are still corroding at a slow rate even after four years exposed to a typical open circulation, aerobic marine environment. There is evidence of aluminium and iron corrosion on all vessels indicating pitting corrosion of the superstructure and steel hull but the anti-corrosive paint system, albeit failing in some areas, is still providing considerable protection to the vessels. The vessels are still in the initial stages of corrosion without encapsulation by concretion, although there has been considerable secondary colonisation of the metal surfaces. The corrosion mechanisms are slowly changing over time and the effects of galvanic corrosion are beginning to be discerned after four years, but the effect is not significant at this stage. The wrecks are successful dive sites and artificial reefs and the impact on local environments has been minimal after five years of submersion, however complex corrosion behaviour is being exhibited by these vessels and it is still too early in the monitoring programmes to make any definitive statements regarding their long-term effect on the marine environment and the stability of the vessels. Therefore, it is of paramount importance that the corrosion and sediment monitoring programmes continue over the next ten to twenty years as the corrosion rates of the vessels will increase significantly over this time, increasing metal enrichment of the sediments and then the long-term environmental effects of these vessels can be properly assessed.

Table 5-1 Monitoring Schedules (Author).

***Swan* scuttling**
14/12/1997

Measurement Timing	Corrosion Survey	Sediment Sampling
Baseline (0)		15/11/1997
0,0	14/12/1997	
5,0.4	16/5/1998	16/5/1998
12,1	30/11/1998	12/12/1998
51,4.3	26-27/3/2002	

***Perth* scuttling**
24/11/2001

Measurement Timing	Corrosion Survey	Sediment Sampling
Baseline (0)		15/11/2001
0,0	26/11/2001	
6,0.5	11-12/6/2002	17/5/2002
16,12&1.3,1	12-14/2/2003	17/12/2002
39,26&3.2,3	8-10/2/2005	16/12/2004
60,5		4/12/2007

***Hobart* scuttling**
15/11/2002

Measurement Timing	Corrosion Survey	Sediment Sampling
Baseline (0)		25/3/2002
2,0.2	28/1/2003	
6,0.5	10/5/2003	
26,2.2	7/1/2005	

Note: Months, years mentioned first correspond to the corrosion survey schedules and those mentioned second correspond to the sediment sampling schedules.

Table 5-2 Corrosion Survey Results for *Swan* (Author).

Point	Metal	Depth (m)	Ecorr (V) Baseline	Ecorr (V) 1 Year	Ecorr (V) 4.3 Years	pH 4.3 years
1	Fe	21.7	0.595	0.506	0.518	8.84
2	Fe	23.7	0.577	0.517	0.514	8.59
3	Fe	22.3	0.597	0.522	0.514	8.61
4	Al	20.9	0.586	0.525	0.558	8.13
5	Al	12.1	0.597	0.527	0.410	7.99
6	Al	17.3	0.563	0.535	0.554	7.64
7	Fe	22.2	0.500	0.524	0.550	7.47
8	Fe	20.3	0.585	0.532	0.564	7.94
9	Al	20.1	0.556	0.522	0.546	8.26
10	Al	21.1	0.522	0.523	0.382	8.15
11	Al	19.7	0.559	0.517	0.368	nd
12	Fe	20.3	0.582	0.508	0.502	8.27
13	Fe	24.2	0.605	0.512	0.368	8.29
14	Fe	21.8	0.599	0.509	0.492	8.30
15	Fe	21.3	0.584	0.508	0.508	8.28

Note: nd = not determined

Table 5-3 Corrosion Survey Results for *Perth* (Author).

Point	Metal	Depth (m)	Ecorr(V) Baseline	Ecorr(V) 0.5 years	Ecorr(V) 1.3 years	pH 1.3years	Ecorr(V) 3.2 years	pH 3.2 years
1	Fe	28.8	-0.494	nd	-0.546	7.97	-0.551	8.26
2	Fe	29	-0.504	nd	-0.531	7.99	-0.556	6.88
3	Fe	29.3	-0.498	nd	-0.535	8.18	-0.554	7.92
4	Fe	28.9	-0.510	-0.508	-0.538	8.15	-0.574	8.11
5	Fe	28.9	-0.538	-0.498	-0.565	7.81	-0.569	7.60
6	Fe	24.8	-0.520	-0.478	-0.525	8.22	-0.546	6.83
7	Fe	23.2	-0.434	-0.480	-0.523	8.17	-0.544	8.21
8	Fe	21.9	-0.504	-0.490	-0.534	8.06	-0.522	8.17
9	Al	26.9	-0.492	-0.500	nd	nd	-0.580	7.92
10	Al	26.8	-0.496	-0.502	-0.554	7.98	-0.576	5.36
11	Fe	27.3	nd	-0.502	-0.556	8.19	-0.583	8.27
12	Fe	26.8	-0.486	-0.502	-0.554	8.19	-0.58	8.25
13	Al	27.4	nd	-0.508	-0.567	8.12	-0.585	8.04
14	Al	27.4	nd	-0.512	-0.564	8.11	-0.588	7.97
15	Al	24.5	nd	-0.514	-0.570	8.05	-0.589	8.22
16	Al	24.3	nd	-0.516	-0.570	8.17	-0.588	8.18
17	Al	23.3	nd	-0.524	-0.579	6.38	-0.600	7.78
18	Al	21.4	nd	-0.524	-0.579	6.38	-0.600	7.78
19								
20								
21	Al	22.5	-0.518	-0.510	-0.565	nd	-0.572	7.71
22	Al	22.4	-0.518	-0.500	-0.552	8.17	-0.572	7.63
23	Al	18.7	-0.518	-0.508	-0.560	8.17	-0.581	7.17
24	Al	13.6	-0.510	-0.526	-0.569	8.16	-0.580	8.19

Note: nd = not determined

Table 5-4 Corrosion Survey Results for *Hobart* (Author).

Point	Metal	Depth (m)	Ecorr(V) 0.2 years	Ecorr(V) 0.5 years	Ecorr(V) 2.2 years	pH 2.2 years
1	Fe	22.5	-0.512	-0.496	-0.536	8.23
2	Fe	22.3	-0.520	-0.503	-0.541	6.80
3	Fe	22.9	-0.521	-0.505	-0.542	5.88
4	Fe	22.1	-0.534	-0.534	-0.581	4.82
5	Fe	21.6	-0.525	-0.523	-0.570	7.65
6	Fe	18.7	-0.494	-0.505	-0.532	8.19
7	Fe	16.8	-0.505	-0.507	-0.537	8.17
8	Fe	15.4	-0.515	-0.513	-0.548	7.70
9	Al	20.3	-0.531	-0.522	-0.574	8.14
10	Al	20.2	-0.535	-0.520	-0.571	8.16
11	Fe	20.3	-0.532	-0.519	-0.575	7.94
12	Fe	20.3	-0.531	-0.520	-0.574	8.04
13	Al	20.1	-0.543	-0.526	-0.578	8.15
14	Al	20.2	-0.542	-0.524	-0.580	8.15
15	Al	17.5	-0.554	-0.536	-0.591	8.01
16	Al	17.5	-0.550	-0.539	-0.585	8.04
17	Al	16.5	-0.561	-0.548	-0.595	8.13
18	Al	13.2	-0.562	-0.568	nd	nd
19						
20						
21	Al	16.1	-0.551	-0.547	-0.587	8.12
22	Al	15.6	-0.534	-0.543	-0.577	8.00
23	Al	11.5	-0.540	-0.544	-0.585	8.10
24	AL	6.5	-0.549	-0.550	-0.582	7.76

Note: nd = not determined

Acknowledgements

The authors wish to acknowledge The Geographe Bay Artificial Reef Society, Dunsborough, Robert Fenn from the City of Albany, Damien Kitto from the South Australian Tourism Commission, Bill Jeffery, Terry Arnott and Rick James from the Heritage Branch of the Department for Environment and Heritage, SA and Ron Moore, Albany Scuba Diving Academy for co-funding our travel, accommodation, analytical costs, logistics, etc. for these monitoring programmes. Last but not least, we would like to thank every staff member from the Department of Materials Conservation and the Department of Maritime Archaeology, Western Australian Museum who have been involved in this project from its inception.

References

ANZECC/ARMCANZ
2000   *Australian and New Zealand Guidelines for Fresh and Marine Water Quality*,
        National Water Quality Management Strategy Paper No 4, Australian and New Zealand
        Environment and Conservation Council and Agriculture and Resource Management
        Council of Australia and New Zealand, Canberra, ACT.

Chant, C.
1984   *Naval Forces of the World*, William Collins Sons & Co, London, England.
Environment Australia
1984   *The Commonwealth Environment Protection (Sea Dumping) Act 1981*, Commonwealth of Australia, Canberra, ACT.
2002   *National Ocean Disposal Guidelines for Dredged Material*, Commonwealth of Australia, Canberra, ACT.
Gillet, R. (editor)
1986   *Australia's Armed Forces of the Eighties*, Child & Henry Publishing, New South Wales.
Gillet, R.
1988   *Australian & New Zealand Warships Since 1946*, Child & Associates Publishing, New South Wales.
Gillet, R., and C. Graham
1977   *Warships of Australia*, Rigby Limited, Sydney, New South Wales.
MacLeod, I. D.
2003   *Metal Corrosion in Chuuk Lagoon: A Survey of Iron Shipwrecks and Aluminium Aircraft*, Report to the US National Parks Authority, Pacific Division, San Francisco, USA, Western Australian Museum, Fremantle, Western Australia.
MacLeod, I., P. Morrison, V. Richards and N. West
2004   Corrosion Monitoring and the Environmental Impact of Decommissioned Naval Vessels as Artificial Reefs, *Metal 04: Proceedings of the International Conference on Metals Conservation, Canberra, 4-8 October 2004,* J. Ashton and D. Hallam, editors, pp. 53-74, National Museum of Australia, Canberra, ACT.
Morrison, P.F.
1998   *Biological Monitoring of the HMAS Swan, 1st Annual Report*, Sinclair Knight Merz, Perth, Western Australia.
2002   *HMAS Hobart Project. Benthic Marine Survey*, Sinclair Knight Merz, Perth, Western Australia.
2003   *Biological Monitoring of the Former HMAS Swan, 5th Annual Report*, Sinclair Knight Merz, Perth, Western Australia.
2007   *HMAS Perth Project. Sediment Monitoring Program – Year 5*, Sinclair Knight Merz, Perth, Western Australia.
Odgers, G.
1989   *Navy Australia, an Illustrated History*, Child & Associates Publishing, New South Wales.
Richards, V.L.
2003   *Corrosion Survey of the Former Naval Vessel HMAS Hobart. May 2003*, Department of Materials Conservation, Western Australian Museum, Fremantle, Western Australia.
Richards, V.L. and I.D. MacLeod
2004   *Corrosion Survey of the Former Naval Vessel HMAS Perth*, Department of Materials Conservation, Western Australian Museum, Fremantle, WA.
Russel, B.C.
1975   Man-made Reef Ecology. A Perspective View, *Proceedings of the First Australian Symposium on Artificial Reefs*, Brisbane, Queensland.

6   Reaching Out to the Community: Bringing Leslie and Ross Back Home to Harcourt

Anne-Louise Muir
Heritage Victoria Conservation Laboratory, 4 Harper Street, Abbotsford, Victoria, 3067

Isa Loo
Western Australian Museum, Department of Materials Conservation, Western Australia
Museum, Shipwreck Galleries, 45 - 47 Cliff Street, Fremantle, Western Australia, 6160

The Heritage Victoria Conservation Laboratory is the only archaeological conservation laboratory in Victoria. When development is proposed on a significant terrestrial archaeological site, a Consent is usually issued with conditions that when an archaeological excavation is conducted an Archaeological Conservation Bond may be levied for the conservation and management of material artefacts. Artefacts are then conserved and the collection managed for the purpose of research, education, publicity and exhibition. Some of the methods that Heritage Victoria are using to present the importance of archaeological material and its conservation to the wider public are public tours, volunteer programs and scientific research. A current collaboration with the Harcourt Historical Society, TerraCulture and VicRoads is the exhibition and interpretation of objects excavated from the Leslie & Ross Railway Construction Camp. This collaboration provides a model for future engagement with communities wanting to connect to their archaeological past.

The Heritage Victoria Conservation Laboratory, as part of the Department of Planning and Community Development (DPCD) is the only conservation and collection facility in Victoria, and perhaps Australia, that is dedicated to archaeological material. While the *Victorian Heritage Act 1995* specifies the Museum of Victoria (now Museum Victoria) as the formal place of lodgement for archaeological relics in Victoria, most archaeological assemblages do not fit within the Museum's collections policy. As a result, the Heritage Victoria Conservation Laboratory has become the default place of lodgement as the *Heritage Act 1995* allows the Executive Director of Heritage Victoria to "otherwise determine" the place of lodgement. As a place of lodgement, the Heritage Victoria Conservation Laboratory also fulfils a number of statutory requirements including Section 128 of the *Heritage Act 1995* which stipulates that, if required by the Executive Director, archaeological relics must be made available for identification or conservation. These statutory requirements are complemented by the issuing of Consents which authorise the disturbance of an historical archaeological site and carry certain conditions which may include the levy of a Conservation Bond for the conservation and curation of archaeological artefacts, subsequently carried out by the Conservation Laboratory.

Conservation Agreements were developed to facilitate the conservation of artefacts from archaeological sites. Issued with Consents, Conservation Agreements outline the provision of conservation and curation services for excavated artefacts. The Consents themselves either stipulate an upfront monetary amount that the applicant has to pay to provide for conservation and curation of the assemblage, or a Consent condition states that the applicant will be liable for the costs of work deemed necessary by Heritage Victoria. The laboratory work is understood to include accessioning, cataloguing and documentation, conservation treatments, collection management and reporting. This process is relatively straight-forward:

artefacts from archaeological projects are delivered to the laboratory by the consultants who excavated them and Conservation Bonds are levied on the site developers to provide the resources for conservation and curation work to take place. Details of this process have been previously described by Heritage Victoria Senior Archaeologist Jeremy Smith (2002).

The Laboratory Process

Presently, the Heritage Victoria Conservation Laboratory takes responsibility for the management of the bulk of archaeological material excavated in the State of Victoria. In order to manage this collection to current museum standards, there are processes that are undertaken to ensure its preservation for the long-term. For laboratory staff, this means not only looking after the physical aspects of the collection such as its accession, location and conservation, but making it accessible to the general public as custodians of Victorian heritage.

Collection Management

Heritage Victoria issued the Archaeological Artefact Management Guidelines in 2002 to specify requirements for the packing and labelling of archaeological assemblages, as well as the provision of a catalogue of all archaeological material by the consultant archaeologist. Once delivered to the laboratory, the archaeological assemblage is inventoried, using the catalogue provided by the consultant. This allows the curatorial officer to verify the location and identification of all material and for the repackaging or relabelling of artefacts as required.

A list of significant artefacts from the site is drawn up by the curatorial officer to identify which artefacts require conservation treatment. This is based mainly on the excavating archaeologist's report and recommendations, but may also include objects assessed as having interpretive potential; objects or object types that are not well-represented across the entire collection and any especially fragile objects. This list details not only the identification and location of an object, but the reason for its inclusion on the list, and a rating from one to three. This rating assists the conservator in making conservation decisions based on object priority given the often limited budget for most projects. The catalogue is checked for compatibility with the current Artefact and Conservation Database, amended where necessary, and is uploaded into the system.

Preventive Conservation

On delivery, the archaeological material is inspected for pest activity. As a precaution and as part of standard museum practice, all organic matter is placed in the laboratory freezer for up to two weeks at -20°C to terminate the life cycle of any pests (Florian 1997:90). Additionally, metals are repacked into ten litre air-tight polypropylene tubs with silica gel to prevent further corrosion by reducing the relative humidity (National Parks Service 1999:1).

Remedial Conservation

Archaeological conservation aims to prevent objects deteriorating once they have been exposed to the atmosphere by excavation and to discover the true nature of the original artefact (Cronyn 1990:1-4). Therefore, conservation needs to be seen as part of the archaeological process. Without it, much archaeological information can be lost or left unexploited (Cronyn 1990:4). While on-site conservation of archaeological material is strongly encouraged and information provided via the Artefact Management Guidelines, the Heritage Victoria Conservation Lab specialises in providing laboratory conservation work where conservators are primarily involved in the technical examination and documentation of the artefacts that are received and their subsequent conservation treatment.

Conservation treatment at the Heritage Victoria Conservation Laboratory adheres to a strong 'minimal intervention' approach and our treatment strategy aims to interfere in the least way possible with the archaeological information. A conservation assessment of the objects is made soon after the assemblage arrives at the conservation lab in its entirety. The conservation assessment takes into account the types of material that the artefacts are composed of and the significance of those objects as it is often not possible to treat all the objects within an assemblage. This allows for different levels of conservation treatment that can be employed across the entire assemblage. Some of these categories, adapted from Cronyn (1990), include:

1. No conservation work – where no conservation work is undertaken by the laboratory except for handling and checking.
2. Minimal conservation work – which includes pest management of organic objects, dewatering bulk metal objects and packing into dessicated storage and repackaging objects into suitable housing.
3. Full conservation work – limited predominantly to objects rated as significant and include photography and documentation, cleaning, active stabilisation and reconstruction (where required).
4. Exhibition conservation – further cleaning and aesthetic treatment required for display on top of full conservation work.

Storage

The collection is stored in an environmentally controlled facility (Figure 6-1), where the temperature is maintained at a constant 20°C ± 2°C with 50% ± 5% relative humidity (RH) as prescribed by conservation research into the most suitable conditions for mixed collections (Thomson 1986:87).

Figure 6-1 Heritage Victoria artefact store (Author).

Making the Collection Available for Research and Access

So what happens to all this material? It is accessioned, treated and stored long term in the storage facility at the Conservation Laboratory. While the majority of the archaeological material stored at the laboratory is undertaken as 'salvage' or 'rescue' archaeology, excavated before the site is destroyed, archaeology and conservation are not done for their own sake – there needs to be some demonstration of a larger public benefit. In an attempt to make the collection more accessible to the community, Heritage Victoria has used a number of different approaches.

Access for Research

The Heritage Victoria collection is not a museum collection. The Conservation Laboratory is not open to the public unless by appointment and there are no permanent or ongoing exhibitions. The main purpose of this collection is for research and it is curated and conserved specifically to make it accessible to students or other researchers who may wish to utilise it. A recent research project that makes use of the archaeological ceramics in the collection was undertaken by Dr Alasdair Brooks to develop a Ceramics Reference Collection (Figure 6-2), which is available to students or consultants who wish to use it to identify historical ceramics. Undergraduate and post-graduate students from archaeology at La Trobe University, Cultural Heritage Studies at Deakin University, and Public History and Conservation at The University of Melbourne have used the collection for their research and publications.

Volunteer Programs

The Laboratory also runs a successful and active volunteer program that can support up to five participants, or more during special projects. These volunteers range from conservation and archaeology students to retirees, and they bring a range of skills to the laboratory. These include box-making, photography, library skills, conservation skills, and artefact labelling.

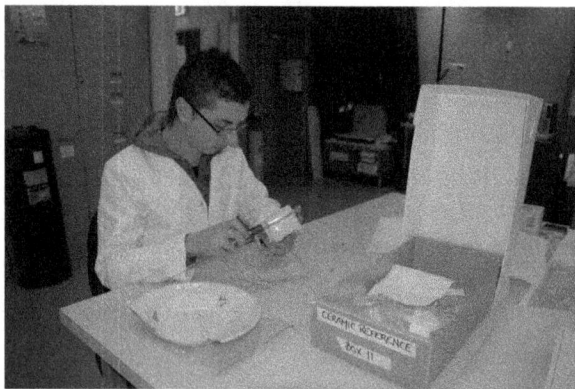

Figure 6-2 Student Intern using Ceramic Reference Collection (Author).

Working under this scheme, Heritage Victoria is able to provide members of the public with opportunities to work with artefacts from land and maritime archaeological sites. Volunteers in the program have worked on archaeological excavations, inventoried the collection, assisted in the installation of exhibitions and supervised simple conservation treatments.

Lab Tours

The Laboratory hosts regular tours of the facility, usually lead by an archaeologist and a conservator. While specific tours have been advertised, sometimes to coincide with events such as National Archaeology Week and National History Week, interested groups have contacted the Laboratory through Heritage Victoria or CAN (Collections Australia Network) websites and requested tours. These groups include metropolitan planners, probus groups, diving groups, catholic archivists, museum guides, Museums Australia members, Australia ICOMOS (International Council on Monuments and Sites) members, students groups, and interested members of the general public. While the tours are unscripted and vary each time, staff explain the differences between terrestrial and maritime archaeology and conservation, talk about various conservation techniques, and allow participants access to objects that are otherwise difficult to make public, such as full wine bottles from shipwrecks which can only be exhibited if kept refrigerated.

The Future of the Collection - Exhibitions and Custodians?

There has been a recent need to focus on the future of the Heritage Victoria collection. This is primarily due to the storage issues currently facing the Conservation Laboratory as a repository. With the increase in Consents being issued in recent years due to a development boom, more archaeological assemblages are being delivered than there is space to store them. To tackle this problem, and also increase the perception of archaeology and its significance to community, there has been discussion on the use of exhibitions and custodians as a way of creating greater public accessibility of the archaeological collection and enhancing the research potential of the collection. The opportunity to participate in a project arising from the excavation of the former Leslie and Ross railway construction camp may have provided a way forward and a model to use on similar cases in the future.

Leslie and Ross

In 2006, the site of the Leslie and Ross construction camp in Central Victoria was excavated by TerraCulture, a Victorian archaeology firm, in response to a VicRoads project to extend the Calder Freeway as part of an upgrade program. The excavation of the Leslie and Ross site by TerraCulture generated a great deal of interest amongst the local community, particularly from the Harcourt Valley Heritage committee. Open days were held and locals were encouraged to volunteer on the excavation (Figure 6-3).

Figure 6-3 Public open day at Leslie and Ross excavation (TerraCulture Pty Ltd).

Leslie and Ross were subcontractors employed to construct part of the railway line between Melbourne and the Murray River. They employed men in a range of different professions to work on the railway. As the majority of them were gold prospectors, their desire to move off to the latest gold strike often hampered construction timetables (Hayes 2008). During the construction of the Elphinstone to Sandhurst (modern Bendigo) section, labourers and their families were stationed near Porcupine Hill where a gravel crushing plant created ballast for the lines. The site was occupied between 1858 and 1863 (Raybould 2006).

Nearby the railway construction camp was the Porcupine Inn, excavated by Vincent Clark and Associates in 1999. Porcupine Inn was one of the earliest pubs on the road from Melbourne to Bendigo, dating from 1846 (Clark 1999:1). It was extremely popular during the Gold Rush, and became a boarding house and watering hole for workers and their families from the Leslie and Ross railway construction site.

Public Accessibility

The Consent issued to VicRoads for the excavation of the Leslie and Ross site included a number of conditions relating to public accessibility to the archaeological heritage of the site. These included:

- A program of site interpretation (including site signage, artefact displays and other forms of interpretation) to be conducted in conjunction with the Harcourt Valley Heritage Committee.

- Careful removal, storage and reinstatement of granite machine base and installation of brass plaque to commemorate the significance of the site.

- All portable relics excavated are to be listed in an Inventory and retained and managed as per the Archaeological Artefacts Management Guidelines.

- Any significant portable relics recovered from the site are to be catalogued, stored and conserved to the satisfaction of the Executive Director, Heritage Victoria.

- After excavation, the objects were lodged at the Heritage Victoria Conservation laboratory.

Nathan Mullane from VicRoads initiated a meeting involving Cathy Tucker from TerraCulture, Heritage Victoria staff, and Neil Charter and George Milford of the Harcourt Valley Heritage Committee to discuss developing an exhibition in fulfilment of consent conditions. The Harcourt Valley Heritage Committee had enthusiastically identified a number of themes that they wanted illustrated in the exhibition, to be installed in the Harcourt Historical Society building. Some of those themes included the industrial work that took place at the site, the domestic lives of the workers, and the archaeological excavation itself.

To this end, Sarah Hayes of TerraCulture worked closely with the Harcourt Valley Heritage Committee on some of the interpretation themes for the exhibition, while Heritage Victoria staff Annie Muir and Isa Loo provided curatorial and conservation advice, including suitable exhibition display materials, optimal environmental conditions and the suitability of objects for display.

The Artefacts

Due to the heavy industrial nature of the site, the majority of the artefacts recovered from the excavation were composed of iron alloys and included nails, cogs, spikes, horseshoes and platform anchors (Figure 6-4).

These iron alloy objects, particularly from the second phase of the excavation, were badly deteriorating with the corroding iron layer delaminating in sections from the surface. Due to the unstable nature of the iron objects retained from the site, and the

fact that very little material of a domestic nature was recovered, it was decided to include the material from the previous excavation of the Porcupine Inn. Although the Porcupine Inn site was not directly related to the 2006 Leslie and Ross Railway Construction Camp excavation, the bottle, ceramics and clay pipes recovered in the 1999 excavation of the Inn illustrate the more human face of the occupation.

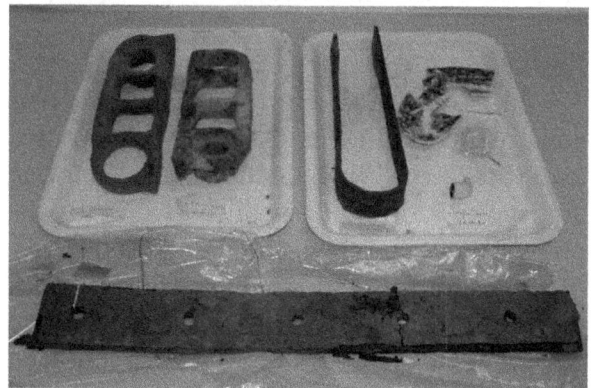

Figure 6-4 Artefacts excavated from the Leslie and Ross site (Author).

Conservation Treatment of Artefacts Selected for Exhibition

The objects selected for display at the Harcourt Valley Heritage Centre were chosen based on a number of criteria agreed upon by the stakeholders. First, the objects had to be robust and able to cope with fluctuating environmental conditions since the Heritage Centre did not have climate control to museum standards. Second, the objects had to fit in the display case purchased by VicRoads to house the exhibition. Finally, they had to complement and illustrate the themes identified by TerraCulture and the Harcourt Valley Heritage Committee. The role of the Heritage Victoria Conservation Laboratory staff was to ensure that the environmental conditions were suited to the selected artefacts for the long-term preservation of these objects, and the treatment of them to exhibition standards.

The conservation treatment of the objects selected for display was limited by time and financial resources. Surface cleaning of the glass and ceramic objects was undertaken to remove ingrained dirt before the metal, objects were tackled. Luckily, most of the iron alloy objects selected for display had previously been treated as part of the whole Leslie and Ross archaeological assemblage and so only minor treatment and preparation was required prior to their installation on site. The iron alloy objects that did require treatment were treated in the same way as the other iron alloy objects which involved using a Volvere dental drill to remove loose surface corrosion products. These were then dewatered before an application of 3% tannic acid was applied and followed by a second dewatering bath in acetone. It was decided to treat the objects with tannic acid even though there would be a substantial colour change to the objects. Given the uncontrolled environmental conditions of the display area, it was thought that further iron corrosion could take place if the active corrosion was not treated. This decision was also influenced by the long-term loan agreement with the Harcourt Valley Heritage Centre, which is a volunteer-based organisation and require a simple process of collection care and maintenance.

Once the conservation treatment and interpretation panels were completed, Isa Loo, from Heritage Victoria and Sarah Hayes from TerraCulture, transported the objects and installed the exhibition at the Harcourt Valley Heritage Centre in Harcourt, Victoria (Figure 6-5).

Custodians

In addition to contributing to the exhibition for the Harcourt community, Heritage Victoria is proposing a future custodian arrangement for some artefacts from the local area. Members of the Harcourt Valley Heritage Committee have expressed considerable interest in a great deal of the material from their region and therefore placing a small portion of it with them will facilitate greater community access to this material and place it within a regional context.

Figure 6-5 Exhibition installed at Harcourt Valley Heritage Centre (Author).

Before a custodial arrangement begins, several factors have to be considered. The ability for the organisation to adequately store the material and to make it accessible to others is important. The Leslie and Ross material was initially considered; however, the size of the assemblage and the metal composition of the majority of objects posed considerable occupational health and safety issues for the Harcourt Valley Heritage Committee. This also created a larger workload for Heritage Victoria staff to monitor the storage conditions and change the silica gel used to control the local humidity of metal objects. The Porcupine Inn assemblage is small and predominantly composed of ceramics and glass which are considered quite stable materials. Although there are a few metal objects, the Porcupine Inn assemblage does not require the same stringent environmental conditions as the Leslie and Ross material. It also contains material that is easier for an historical society to exhibit. Ceramics, glass, clay pipes and other small finds can often be displayed in a more evocative way than large quantities of metal spikes.

Other pressing issues yet to be resolved include the resourcing of preparatory work on the assemblage before delivery. A minimal amount of preparatory work on the assemblage by Heritage Victoria staff such as numbering objects, ensuring suitable packaging and documentation needs to be carried out before the assemblage can be handed over. Whether the assemblage will need to be monitored and maintained while at Harcourt is also an issue to be resolved.

Availability for researchers during custodianship is also an important consideration. The significance of the collection held by Heritage Victoria lies largely in its research value. As such, there is a need to make sure that all assemblages are accessible for appropriate research.

The ability of the Society to adequately store the archaeological material needs to be assessed. Many historical societies lack space and funding. Harcourt, however, is fortunate in that it has a small amount of dedicated storage space.

Conclusion

This collaboration demonstrates a way forward and will provide Heritage Victoria with a model for future engagement with communities in order to reconnect them with their archaeological past and where Heritage Victoria can provide assistance in their long term preservation.

Acknowledgements

The authors would like to thank Cathy Tucker and Sarah Hayes of TerraCulture, Nathan Mullane of VicRoads, Neil Charter and George Milford of the Harcourt Valley Heritage Centre, for help on the project and this paper. We also thank Sheldon Teare for allowing us to include their photos.

References

Clark, V.A.
1999   *The Porcupine Inn (Victorian Heritage Inventory H7724-296): A report on Preliminary Archaeological Investigations*, Report to VicRoads from Dr Vincent A Clark and Associates.
2003   Leslie and Ross Railway Construction Camp Site (Victorian Heritage Inventory H7724-0290), Report to VicRoads from Vincent Clark and Associates.
Cronyn, J.M.
1990   *Elements of Archaeological Conservation*, Routledge, London, United Kingdom.
Florian, M.L.
1997   *Heritage Eaters: Insects and Fungi in Heritage Collections*, James and James, United Kingdom.

Hayes, S.

2008   Leslie and Ross Railway Construction Camp, Exhibition interpretation panels, TerraCulture Pty Ltd.

Heritage Victoria

2002   Archaeological Artefact Management Guidelines, Department of Planning and Community Development.

National Park Service

1999   Using silica gel in Microenvironments, Conserve O Gram 1/8, http://www.nps.gov/history/museum/publications/conserveogram/01-08.pdf

Raybould, O.

2006   *Leslie and Ross Railway Construction Sites, Harcourt (H7724-0290): Archaeological Salvage Excavation*, Report to VicRoads from TerraCulture Pty Ltd.

Smith, J.

2002   Funds for Finds: Heritage Victoria's Artefact Conservation Bond Scheme, *Australian Association of Consulting Archaeologists Newsletter*, 91: 13-15.

State Government of Victoria

1995   *Victorian Heritage Act*, Melbourne, Victoria.

Thomson, G.

1978   *The Museum Environment*, Butterworths, London, United Kingdom.

7 Excavation and Relocation of the Former Hovell Pile Light

Jason Raupp
Department of Archaeology, Flinders University, GPO Box 2100, Adelaide, South Australia, 5001

Cosmos Coroneos
Cosmos Archaeology Pty. Ltd., Maroubra, New South Wales

Jennifer McKinnon
Department of Archaeology, Flinders University, GPO Box 2100, Adelaide, South Australia, 5001

As one of a large number of mitigation responses to the Victorian Government's Port Melbourne Channel Deepening Project, historical and archaeological investigations of the *South Channel UNID Dromana* (H7821-0128) site located in Port Phillip Bay were undertaken by Cosmos Archaeology Pty Ltd. These investigations determined that the submerged structure represented the partial remains of the former Hovell Pile Light (fHPL) (1924-38) and that they would be impacted by dredging operations. As such, an archaeological excavation, recovery and reburial project was designed to best protect the archaeological fabric of the fHPL remains. This paper reports on the identification, excavation, relocation and preservation of the site and highlights a rare form of cultural resource management within Australia, that is, the complete recovery and reburial of a submerged archaeological site.

Introduction

In 2005 a submerged archaeological site located to the north east of the current Hovel Pile Light was reported to Heritage Victoria. Initial observations identified it as a shipwreck and the site was therefore listed on the Victorian Heritage Register as the *South Channel UNID Dromana* (H7821-0128). The site was further investigated as part of the environmental assessment studies associated with the Victorian Government's Port Melbourne Channel Deepening Project. These investigations determined that the site most likely represented the partial remains of the Former Hovell Pile Light (fHPL) (1924-38) and that they would be severely impacted by proposed dredging operations. It was therefore determined that the site should be excavated and large timber components relocated. As such, an archaeological excavation, recovery and reburial project was designed to best protect the archaeological fabric of the fHPL remains. This paper reports on the identification, excavation, relocation and preservation of the site.

Site Environment

The remains of the fHPL were located approximately 1.5 nm north east of the present Hovell Pile Light (Figure 7-1) in Port Phillip Bay. The depth of the site was approximately 16 m and the centre of the site was situated atop a low circular mound approximately 20 m in diameter. The exposed remains of the site protruded from the top of the mound, which rose approximately 0.5 m above the seabed. Timber, metal (ferrous and copper alloy) and glass materials were visible on the site. The bulk of the visible cultural material covered an area of approximately 10 m by 7 m. The seabed around the site was composed of fine-grained sediments (more silt than sand) mixed with dead scallop and mussel shell covering a hard marl seabed.

Previous Archaeological Work

The remains of a possible archaeological site located to the north east of Hovell Pile were reported to HV in 2005 by Southern Ocean Exploration (SOE) divers. While diving on a GPS location provided by a local commercial fisherman, they discovered a row of timber frames protruding upright from a low mound which was surrounded by a jumble of other timbers (Taylor 2007).

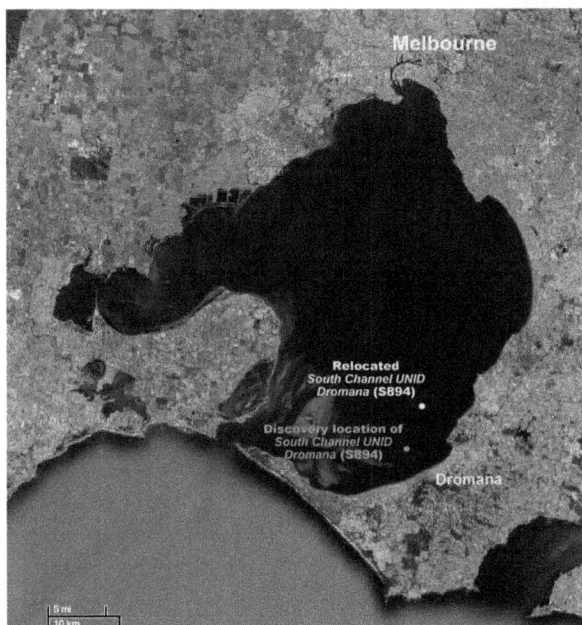

Figure 7-1 Location of South Channel UNID Dromana site (Author).

From 2005 to 2006, SOE undertook site surveys and research aimed at identifying the site's form, function and date of formation. The survey provided a more coherent picture of the site, identifying gas cylinders and the elements of a navigation light system. The basic appearance of the gas cylinders found on the site pointed to an approximate date between the late nineteenth and early twentieth century. The principal observable components of the navigation light were a lantern and copper alloy 'light chimney.'

Unfortunately, the 'light chimney' was looted from the site sometime in 2006 and its theft represents the first of three identified acts of looting and vandalism on the site (Taylor 2007). Subsequent acts of vandalism occurred during and after the excavation of the site. This severely curtailed the archaeologists' ability to interpret the site.

Research Design

The investigation of the site was focused on determining whether the site was actually the remains of the fHPL destroyed in 1938. Furthermore if the site was that of the fHPL, obtaining information on how it functioned was seen as most important as comparisons could be made with other gas-lit navigation aids in Port Phillip Bay.

Site Recording

Trilateration was chosen as the best practical survey methodology for obtaining accurate positional data. Nine datums were positioned around the fHPL. Measurements were taken from at least three of the datums to the same point on each of the recorded timbers and artefacts.

For interpretation purposes, the site was divided into two sections called "topside" which included the hut structure and "underside" which included the platform structure. An identification code was assigned to each timber as they were exposed.

Once assigned, locations were recorded on the site plan and small aluminium, and later white plastic tags were engraved with the project and provenance code. On the following dives these tags were attached to the corresponding timber using galvanized nails. These tags helped divers to navigate and record the site, and were especially beneficial in periods of low visibility.

Photographic and video surveys were undertaken prior to disturbance of the site and when new elements of the site were exposed. Multi-beam sweeps carried out over the site during the excavation by Port of Melbourne Corporation were useful in giving a gross impression of the layout of the site as well as providing accurate data on relative heights of structural features and depths of excavation.

Excavation

The excavation strategy for the Hovell Pile Light investigation was devised prior to commencement, but was refined to increase efficiency as the project progressed. Overburden removal was carried out by Professional Diving Services (PDS) divers on surface supplied diving systems using an airlift. These systems allowed for remote monitoring of dredging operations and conditions by the dive supervisor and principal archaeologist. All detailed excavation was conducted by archaeologists assisted by PDS divers.

The entire topside section of the structure was assessed to have a high potential for containing preserved superstructure and artefacts related to the function of the light and equipment employed (Figure 7-2). The deck and underside sections represented the basic structure of the platform and so were assigned a lower priority due to a reduced risk of damaging sensitive archaeological materials. This assessment allowed for a faster rate of excavation.

Part of the excavation methodology included the assessment and preparation of the four acetylene gas cylinders for lifting. This task was completed by PDS divers and involved deconcretion in order to free them from the structure and to access tank valves. Once accessed, the tank valves were opened

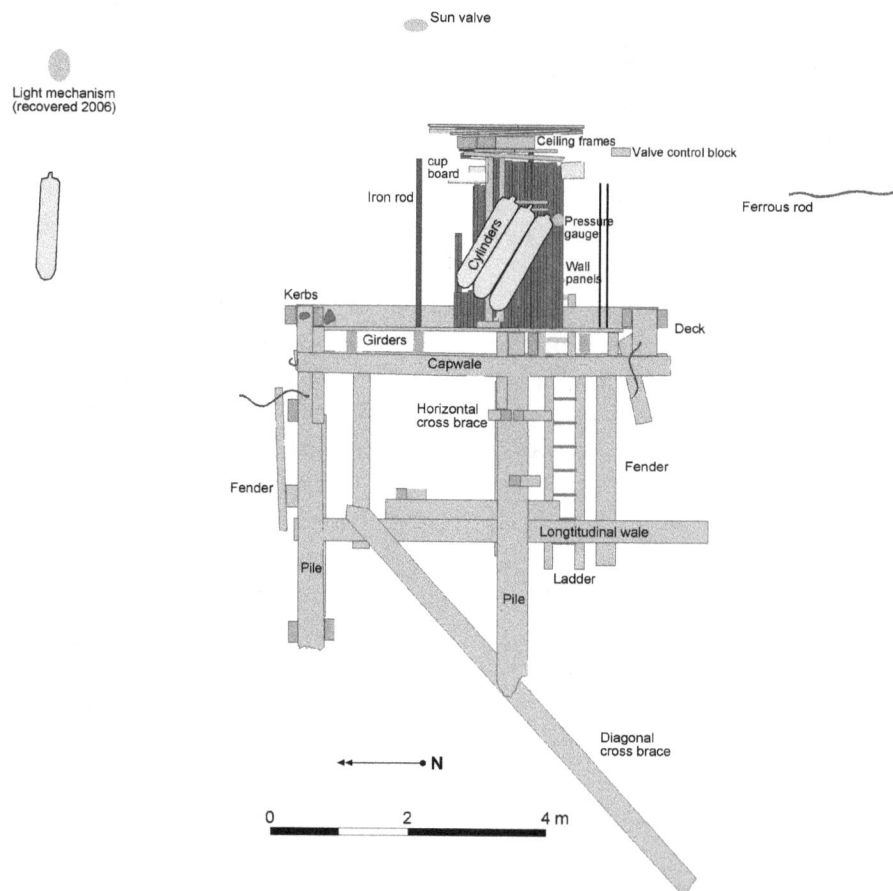

Figure 7-2 fHPL site excavation with some structural interpretation added (Author).

slightly to allow for the release of any residual gas.

Artefact Recording and Recovery

All artefacts observed on the site were recorded, photographed and recovered. Area designations and trilateration information for all artefacts was recorded. To allow uninterrupted excavation, large artefacts, including loose timber elements and the gas cylinders, were recorded *in situ* and (when necessary) moved to a designated off-site recovery point. Prior to removal from the site, artefact numbers and area locations were written on small Mylar tags, which were either included with each object in a separate container (plastic, Hessian bags or plastic boxes) or attached using cable ties. To reduce the risk of damage in transit to the surface, when possible, artefacts were placed inside weighted artefact transport receptacles (plastic chicken crates) with lids.

Once on the surface, all artefacts were measured, described and photographed. In some cases limited removal of silt from the artefacts was conducted for the purposes of obtaining clearer photographs. Artefacts were relabelled with aluminium or white plastic tags (depending on the composition of the artefact) and kept in covered tubs with a 50/50 fresh/salt water solution. Selected artefacts were chosen for further conservation, permanent storage and eventual display. These artefacts were selected based on their condition, suitability for display, and representativeness of the site and/or uniqueness.

Artefacts Recovered

The term artefact is used here to represent both individual objects and groups of objects that were recovered together and are associated with a particular function. A total of 57 artefacts representing different aspects of the light's operation or post-depositional activities were recovered from the fHPL site. In order to simplify this discussion, artefacts have been grouped according to their function. These groupings are:

- *Navigational*: relating to the lighting system and including all components of the lantern, valves and fuel system. A total of 31 artefacts were classified as belonging to the navigational group. Each of these is a component of the acetylene-fuelled system that powered the beacon. The components of this system include cylinders and cylinder valves, copper alloy tubing, a central control block and pressure gauge, mounting straps, the beacon and a solar powered switch called a Sun Valve.
- *Structural*: relating to components of the topside structure. A total of 10 artefacts were classified as belonging to the structural group. These were interpreted to be components of a hut which housed the gas cylinders. The components of this hut include small timbers, plate glass, hinges and fasteners.
- *Recreational*: relating to post-depositional activities on the site. Two of the artefacts recovered were classified as recreational objects. Both of these are cast-lead fishing weights which are triangular in shape and have a hole near their apex through which fishing line is attached. These artefacts may or may not have historical significance as they could be from modern fishing episodes.
- *Unknown*: objects of indeterminate function. Nine of the recovered artefacts were classified as unknown. Most of these objects were heavily concreted and may be better understood with proper cleaning and conservation treatments.

Of the artefacts recovered, only those included in the navigational group provided solid diagnostic evidence.

Raising and Recording the Structure above Water

Once the site had been thoroughly recorded, additional excavation took place around the structure to counteract the effects of suction when the lift took place. PDS divers attached lifting straps to the intact section of the pile platform structure, which were in turn connected to a lift line from the crane aboard a jack up barge positioned adjacent to the site. The intact pile structure was lifted onto the deck of the barge.

That the underside of the structure, the surviving deck platform and piles, came up intact defied predictions; the whole process of recording on the seabed was carried out in the expectation that the site would come up in pieces. This was because the structure was not densely fastened and those iron bolts observed were heavily corroded. This unexpected development allowed further recording of the construction methods and arrangement of timbers, as well as verification of measurements.

Light artefacts, primarily timbers which had been previously bundled on the seafloor, were also retrieved. Loose artefacts exposed when the main structure was lifted were placed in polymer bags and bought to the surface along with the gas cylinders. Each of these artefacts was recorded in detail as most of them were related to the super structure including the hut.

The cylinders were assessed to be unstable as gas (presumably acetylene) was observed escaping through their necks and welds. As a result they were not retained for conservation.

Reconstruction from Archaeological Data

To date no archival information about the construction of the fHPL has been located. Therefore the archaeological data recovered during this project is the only source of information available for the virtual reconstruction of the Pile Light at the time of it destruction (Figure 7-3).

The underside, of the pile structure was of a standard form and construction typical in Port Phillip Bay for pile lights and pier construction in general (Barnard 2008). Nine piles, arrayed in three rows of three, supported the platform. The arrangement was strengthened by longitudinal wales, and horizontal and diagonal cross braces. Vertical fenders ran along the outside edge of the structure and were attached to the piles. Checked into the tops of the piles were cap wales, which supported girders (or joists). Deck planking was in turn fastened to the top of the girders. On each of the four corners of the structure, short sections of timber were checked into the top of the piles. Attached to these kerb supports were the timber kerbs or 'kick boards'. A thick timber ladder with iron rungs was fastened to one side of the structure. It appears that the original ladder was not long enough and so it was lengthened by fastening short timber sections and rungs to the upper part of the platform.

The appearance of the hut structure, which housed the acetylene cylinders and ancillary equipment, is difficult to interpret given the fragmentary remains. This situation is exacerbated by vandalism that took place after the excavation and destroyed much of this section of the site. The angle of the corner frames and wall panels (110 to 115 degrees) indicates that the hut was hexagonal. Within the hut there was a cupboard at head height and there was probably a small window built into the wall. There was no evidence of the wall frames being securely fastened to the deck.

81

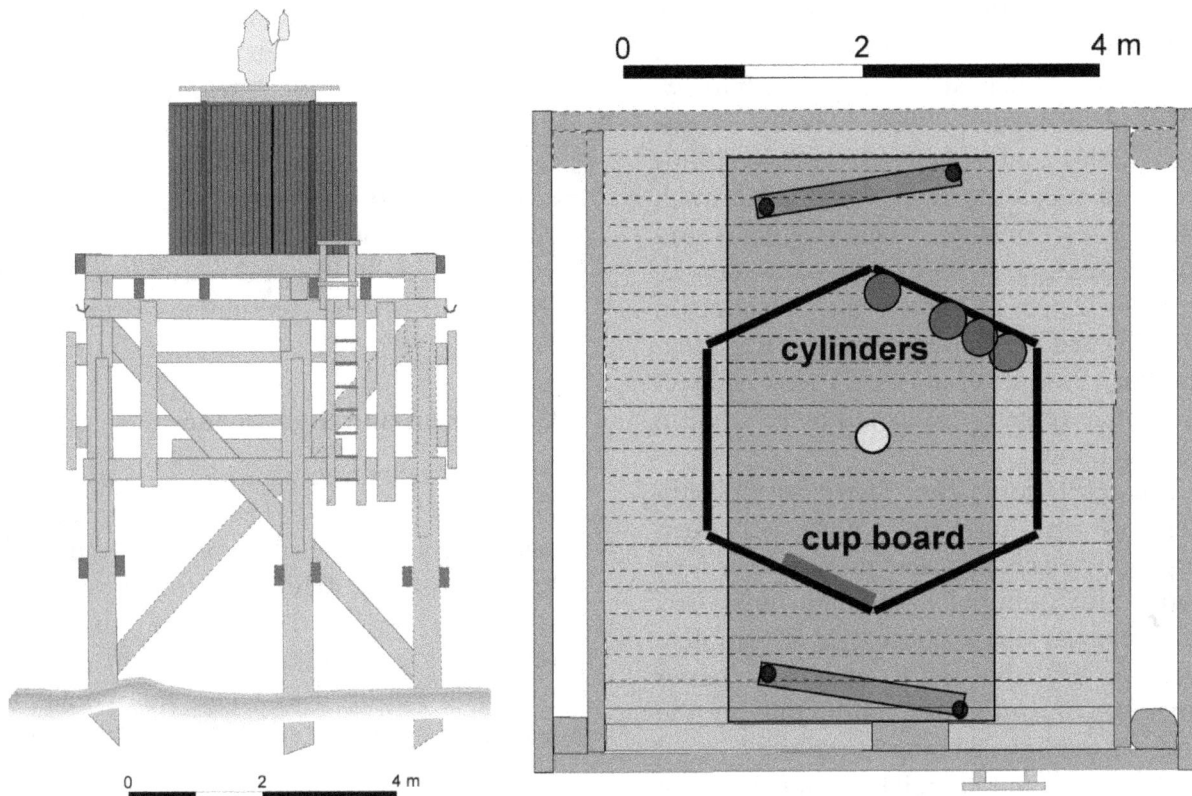

Figure 7-3 Reconstruction of the fHPL based on archaeological evidence (Author).

The arrangement of the ceiling/roof of the hut was also difficult to reconstruct due to the incomplete state of the remains. Timber panelling similar to that used for the hut walls formed the ceiling. The thick frames situated above the ceiling were attached to iron stanchions, which were firmly fixed to the deck. These rods were not built into the walls of the hut but were positioned outside. It appears likely that the hut was fitted under a level and thick timber roof, supported by iron rods. It was upon this roof or 'upper deck' that the pile light's lantern was probably affixed, connected to its fuel source through a hole in the roof/deck and the ceiling.

Relocation and Reburial

In situ preservation through the methods of reburial or re-covering is based on the idea that certain environments are capable of slowing deterioration, such as the anaerobic environment. The goal of reburial is to "re-create a stable environment, slowing chemical, biological and physical deterioration" (Ortmann et al. 2009: 6). This can be accomplished through a number of ways either through covering the site with barriers (i.e. sand bags) or geotextiles (i.e. debris netting or artificial sea grass) to create sediment deposition, or through sediment drops and backfilling techniques.

It is pertinent here to make a distinction between backfilling which is a normal process of archaeological excavation on dry land or underwater (leaving as site as it was found), and reburial which is an active process of adding additional sediment to a site. Most shipwrecks in Australia have been subject to backfilling methods after excavation rather than the latter technique of reburial. For example, the site of *Pandora* was actively backfilled after excavation

82

(Guthrie et al. 1994; Gesner 1993) as was the *James Matthews* (Nyström Godfrey et al. 2005) site. Other sites such as *William Salthouse* (Steyne 2009), *Sydney Cove* (Nash 2006) and *Solway* (Coroneos 1996) have been subject to active reburial through dumping sediments, sandbagging and artificial sea grass. Still other sites such as the *Day Dawn* (Kimpton and Henderson 1991: 25) and the shipwreck site at Red Bay, Labrador (Stewart et al. 1995) were subject to complete recovery followed by reburial in a new location.

The fHPL joins *Day Dawn* as the only other site in Australia which has been completely excavated and moved to a new location for reburial. Although, the *Day Dawn* reburial was unsuccessful at complete coverage and the fHPL stands a better chance at *in situ* preservation due to the metres of sand placed atop of the site.

The fHPL and associated artefacts were deposited in the South East Dredged Material Ground (DMG) at a precise designated positioned, approximately 7 km to the north east. The decision to relocate the remains of the site within the DMG was based on the assessment that they would become buried with up to 4.5 m of dredged sands, hence preserving the cultural material in an anaerobic environment. While no monitoring is planned for the site, it should be noted that sediments could erode or be further deposited on site. As stated the position of the redeposited site has been accurately recorded should plans for future excavation or monitoring eventuate.

The water depth at the redeposit location is approximately 19.5 m. This exceeded the depth at which the jack up barge could safely drop its spuds and operate its crane. As a result, the intact structure of the pile light was pushed over the side of the barge rather than being lowered via crane. It was not considered feasible or necessary to cover the structure in plastic or some other protective layer as sand was to be deposited on the site shortly after its relocation. After the remains were deposited, divers observed that the structure remained largely intact on the seabed. Loose timbers and smaller artefacts not deem significant for conservation were tied into bundles and within the cylinders and were placed within the body of the remaining structure by divers.

Conclusion and Significance

The excavation of *South Channel UNID Dromana* (H7821-0128) confirmed the assessment that the site represents the partial remains of the fHPL, destroyed during a storm in April 1938. The excavation revealed a conventional pile light structure for Port Phillip Bay. The hut containing acetylene cylinders and a mount for the light apparatus is similar to other pile lights through its hexagonal shape. The light itself seems to have been supported by a timber platform raised above the hut by iron stanchions. The hut appears to have been lightly constructed which supports the idea that it did not support the full weight of the light. The lighting system appears to have been standard for the time, with welded pressure cylinders, coiled copper tubing, a gauge, control block, sub valve, the lantern itself and the lantern hood.

The significance of the fHPL (1924-1938) lies in its ability to increase our understanding of the evolution of navigation aids and maritime infrastructure generally in Port Phillip Bay - especially at a time when the fuel source for pile lights was being converted from kerosene to acetylene. The site is important in this respect as it has not been altered, upgraded or replaced unlike existing pile lights. It may be possible that records relating to this structure are found in archival sources, including places such as the Queenscliffe Maritime Museum. Such finds would only enhance and augment the information that was recovered from the archaeological investigation of this site.

The project itself is significant as it is a successful response to seabed development, in this case capital dredging. Government agencies, commercial divers and professional archaeologists worked closely together to record and archaeologically excavate the site, recover artefacts for conservation, recover the intact remains and rebury them in a new location. To date this is a unique event within an Australian context as it represents the complete recovery of a site, relocation and reburial for *in situ* preservation. Although no monitoring project was put into place, it is anticipated that the site will be preserved under metres of sand for years to come.

Acknowledgments

The authors would like to thank Malcolm Venturoni and the staff of PDS. Special thanks go to archaeologists Caroline Wilby, Shaun Mackay and Kenny Keeping. Final thanks go to Heritage Victoria (Peter Harvey, Cassandra Philippou and Hannah Steyne) and Port of Melbourne Corporation (Geoff Nicol) for laying down a trouble free and painless process throughout all phases of the project.

References

Barnard, J.

2008 *Jetties and Piers: A Background History of Maritime Infrastructure in Victoria*, Melbourne, Heritage Council of Victoria.

Corones, C.

1996 The *Solway* (1837): Results of the 1994 Test Excavation, *Bulletin of the Australian Institute for Maritime Archaeology* 20(1): 19-38.

Gesner, P.

1993 Managing *Pandora*'s Box – the 1993 *Pandora* Expedition, *Bulletin of the Australian Institute for Maritime Archaeology* 17(2): 7-10.

Guthrie, J.N., L.L. Blackall, D.J.W. Moriarty and P. Gesner

1994 Wrecks and Marine Microbiology: Case Study from the *Pandora*, *Bulletin of the Australian Institute for Maritime Archaeology* 18(2): 19-24.

Kimpton, G. and G. Henderson

1991 The Last Voyage of the *Day Dawn* Wreck, *Bulletin of the Australian Institute for Maritime Archaeology* 15(2): 25-28.

Nash, M.

2006 Individual Shipwreck Site Case Studies, In *Maritime Archaeology: Australian Approaches*, M. Staniforth and M. Nash, editors, pp. 55-67, Springer Press.

Nyström Godfrey, I., I. Jakubowicz, N. Yarahmadi, S. Ekendahl, and SP Swedish National Testing and Research Institute

2007 Appendix 6: Investigation of the Effects of Burial on Materials Used At Archaeological Excavations to Pack and Label Objects - Final Report, Phase 1, In *Reburial and Analyses of Archaeological Remains: Studies On the Effects of Reburial On Archaeological Materials Performed at Marstrand, Sweden 2002-2005. The RAAR project T*, Bergstrand and I. Nyström Godfrey, editors, pp.2-51, Bohusläns Museum and Studio Västsvensk Konservering, Helsinki.

Ortmann, N., V. Richards, and J. McKinnon

2009   *In Situ* Maritime Preservation and Storage in Maritime Archaeology – Part 1. Draft under review *International Journal of Nautical Archaeology.*

Stewart, J., L.D. Murdock, and P. Waddell

1995   Reburial of the Red Bay Wreck as a Form of Preservation and Protection of the Historic Resource, *In Material Issues in Art and Archaeology: IV*, P.V. Vandiver, J.R. Druzik, J.L.G. Madrid, I.C. Freestone and G.S. Wheeler, editors, pp. 791-806, Materials Research Society, Pittsburgh.

Steyne, H.

2009   Cegrass™, Sand & Marine Habitats: A Sustainable Future for the *William Salthouse.* In In Situ *Conservation of Cultural Heritage: Public, Professionals and Preservation,*V. Richards and J. McKinnon, editors, pp.40-49, Past Foundation.

Taylor, P.

2007   *Light Project 2006*, Report to Heritage Victoria from Southern Ocean Exploration.